The Air Pilot's **Manual**

Volume 2

Aviation Law, Flight Rules and **Operational Procedures**

Meteorology

Revised & edited by
Peter D. Godwin

'Recommended reading'
Civil Aviation Authority LASORS

Air Pilot Publishing

Copyright © 1987-2004 Aviation Theory Centre

ISBN 1 84336 066 7

First edition published 1987
Second revised edition 1987
Reprinted 1988
Reprinted with amendments 1989
Reprinted 1990
Third revised edition published 1993
Reprinted with amendments 1994
Reprinted twice with amendments 1995
Fourth revised edition published 1997
Reprinted with amendments 1998
Fifth revised edition 1999
Sixth edition 2001
Reprinted with amendments 2002
Seventh edition 2004

Origination by Bookworks Ltd, Ireland.

Printed in England by Stanley L Hunt (Printers) Ltd, Rushden, Northants.

Published by Air Pilot Publishing Ltd
Mill Road, Cranfield
Beds MK43 0JG England
Website: www.pooleys.com
Tel: (01234) 750677
Fax: (01234) 750706

The Air Pilot's **Manual**

Volume 2

Contents

Exercises and Answers

JAR-FCL Abbreviations

Index

Editorial Team

Peter Godwin

Head of Training at Bonus Aviation, Cranfield (formerly Leavesden Flight Centre), Peter has amassed over 14,000 instructional flying hours as a fixed-wing and helicopter instructor. He has edited this series since 1995 and recently updated it to cover the JAR-FCL. As a member of the CAA Panel of Examiners, he is a CAA Flight Examiner for the Private Pilot's Licence (FEPPL[A]), Commercial Pilot's Licence (FECPL[A]), Flight Instructor Examiner (FIE[A]), as well as an Instrument Rating and Class Rating Examiner. A Fellow of the Royal Institute of Navigation (FRIN), Peter is currently training flying instructors and applicants for the Commercial Pilot's Licence and Instrument Rating. He has been Vice Chairman and subsequently Chairman of the Flight Training Committee on behalf of the General Aviation Manufacturers' and Traders' Association (GAMTA) since 1992 and is a regular lecturer at AOPA Flight Instructor seminars. In 1999 Peter was awarded the Pike Trophy by The Guild of Air Pilots and Air Navigators for his contribution to the maintenance of high standards of flying instruction and flight safety. Previously he was Chief Pilot for an air charter company and Chief Instructor for the Cabair group of companies based at Denham and Elstree.

Dorothy Pooley LLB(Hons) FRAeS

Dorothy holds an ATPL and is both an instructor and examiner, running Flight Instructor courses at Shoreham and she is Training Captain for an AOC operator also at Shoreham. She is also a CAA Flight Instructor Examiner. In addition, having qualified as a solicitor in 1982, Dorothy acts as a consultant to ASB Law, where she specialises in aviation and insurance liability issues. She has lectured widely on air law and insurance issues. This highly unusual combination of qualifications has led to her appointment as Honorary Solicitor to the Guild of Air Pilots and Navigators (GAPAN). A Fellow of the Royal Aeronautical Society, she chairs the GAPAN Instructor Committee, and is on the Education & Training Committee, as well as serving as an Assistant on their Court. In 2003 Dorothy was awarded the Jean Lennox Bird Trophy for her contribution to aviation and support of Women in Aviation and the BWPA (British Women Pilots Association). A regular contributor to seminars and conferences, Dorothy is the author and editor of a number of flying training books and has published articles in legal and insurance journals.

Graeme Carne

Graeme is a BAe 146 Captain with a dynamic and growing UK regional airline. He has been a Training Captain on the Shorts 360 and flew a King Air for a private company. He learned to fly in Australia and has an extensive background as a flying instructor in the UK. He has also been involved in the introduction of JAR OPS procedures to his airline.

Bill Ryall

A highly experienced pilot, Bill is currently an editor of *Pooley's Flight Guide,* having retired from the RAF after 38 years' service. He flew Lancasters during the war, followed by a spell on transport aircraft and twelve years with Overseas Ferry Command, during which he flew a great variety of aircraft, from Dakotas to twin jets. Later, he became a VIP pilot and an Instrument Rating Examiner, and was decorated for distinguished flying and awarded the QCVSA. Bill's experience was broadened by tours of duty as an APP/Radar Controller and on the editorial staff of No. 1 AIDU Northolt. He still holds a pilot's licence and flies regularly.

Robert Johnson

Bob produced the first three editions of this manual. His aviation experience includes flying a Cessna Citation II-SP executive jet, a DC-3 (Dakota) and light aircraft as Chief Pilot for an international university based in Switzerland, and seven years on Fokker F27, Lockheed Electra and McDonnell Douglas DC-9 airliners. Prior to this he was an Air Taxi Pilot and also gained technical experience as a Draughtsman on airborne mineral survey work in Australia.

Warren Yeates

Warren has been involved with editing, indexing, desktop publishing and printing flight training manuals since 1988 for the UK, US and Australian markets. He currently runs a publishing services company in Ireland.

Acknowledgements

The Civil Aviation Authority, Captain R. W. K. Snell (CAA Flight Examiner [ret.]), Peter Grant, John Monroe, Marian Ashby, Aidan Bell, Stephen Burt, the late John Fenton, Shaun McConnell, Edward Pape, Robert Pooley, and the many other instructors and students whose comments have helped to improve this manual.

Preface to the 7th Edition

So many changes have taken place in the last two years, necessitating legislative updates and the consequential implementation into practice throughout the flight training industry, that it has been necessary to produce a new edition. This edition completely updates both the aviation law and meteorology sections and takes account of the widespread availability of briefing materials for pilots via the Internet.

As before, unique in PPL training manuals, you will find throughout the text comprehensive references to the source documents and relative section numbers. This will enable you to carry out further research should you wish to delve deeper than the PPL syllabus.

To highlight the difference between UK CAA material and ICAO or JAR material, we use a lighter typeface for the ICAO material.

Section **One**

Aviation Law, Flight Rules
and **Operational Procedures**

Aviation Law and Legislation

Introduction

It is a fact of life that most activities we undertake, whether for leisure or work, are regulated to some degree, very often with safety being a prime purpose. As you learn to fly and gain wider experience you will encounter many types of rules and procedures, governing such things as the operation of aircraft, types of airspace, licence privileges and Rules of the Air.

In order to set your study of aviation law in context it is useful to understand where the rules and regulations are to be found and how they relate to each other.

Almost since aviation began, the rules and regulations have developed on an international basis.

International Civil Aviation Organization (ICAO)

To highlight the difference between UK CAA material and ICAO or JAR material, we use a lighter typeface for the ICAO material (as opposite).

After the First World War ended, civil flying resumed in 1919 and, with four years of accelerated development during the war, aircraft performance had improved immeasurably. The benefits of air travel to facilitate the rapid crossing of international boundaries and to communicate with distant nations for transportation and trade were realised. However, many problems had to be overcome, such as overflying sovereign territories of different nations and landing safely at prepared airfields.

The Paris Convention of 1919 was the first major international meeting of aviation-oriented nations, though it was more a European than a worldwide gathering. Nonetheless, general principles of air law and aviation procedures were agreed and adopted.

Further conferences followed until the historic Chicago Convention in 1944 placed a moral obligation on contracting States (nations) to provide safe and efficient ground and flight organisations in their territories for the development of international aviation. Following the entry into force of the Chicago Convention, a permanent International Civil Aviation Organization was formed in 1947. It was based in Montreal, Canada, where it remains to this day. The purpose of ICAO is to promote aviation standards and recommended practices internationally.

The United Kingdom, like most nations of the world, is an active member of ICAO.

Standardisation has occurred to a remarkable degree since the signing of the *Chicago Convention* in 1944, although, as is always the case in international affairs, some differences remain.

For example:

☐ Cruising levels in Western nations are based on **feet**, whereas in some Eastern European countries they are based on **metres**.

☐ Many countries now use the **hectopascal (hPa)** as the unit for pressure (on ICAO recommendation), whereas the UK has decided to retain the **millibar (mb)** unit for the foreseeable future (although the units represent the same amount of pressure, i.e. 1 hPa = 1 mb). The United States uses inches of mercury.

☐ The airspace classification system recently introduced in the UK is also being adopted by most countries; however, there are small differences in how the system is being implemented in each country, e.g. Class B is allocated to Upper Airspace in the UK, whereas in the US system, Class B is being allocated to Terminal Control Areas (and the descriptive term for the airspace is being dropped). The specifications of ATC service, VMC minima, clearance requirements, radio equipment, etc. for each airspace class are much the same.

However, these differences do not detract from the fact that basic flight rules and procedures are similar throughout the world.

Chicago Convention

The broad principles of the ICAO as laid down by the Chicago Convention are known as **Articles**. The text of these Articles is set out in the ICAO document known as DOC 7300 and is amended when necessary. Implementation is the responsibility of the contracting State.

DOC 7300 contains the text of the Convention on International Civil Aviation as signed at Chicago in 1944. A summary of some of its 96 Articles applicable to student and private pilots follows. You need to become familiar with these articles because they form part of the PPL syllabus and are included in examinations. The term *State* denotes a contracting nation to the Chicago Convention.

Article 1 – Sovereignty

Each State (nation) that has signed the Convention is recognised to have complete and exclusive sovereignty over the airspace above its territory.

Article 2 – Territory

A State's territory is the land and territorial waters over which the State has sovereignty.

Article 4 – Misuse of civil aviation

Each State agrees not to use civil aviation for any purpose contrary to the aims of the Convention (i.e. for illegal purposes or for war).

*The words **shall** and **should** have precise meanings in ICAO documents. For instance, where it says a pilot **shall** do something, it means he must, he is required to. Where it says a pilot **should** do something, it means he doesn't have to, but in the interests of good airmanship and flight safety it is recommended that he does.*

Article 5 – Right of non-scheduled flight

Each State will allow aircraft from all other States (except for scheduled international flights) to fly into or through its airspace, and to land without prior permission. States also have the right to require overflying aircraft to land. Where terrain is remote and navigation facilities are inadequate, States may require overflying aircraft to follow prescribed routes, or to obtain prior permission for the flight.

Article 10 – Landing at customs airport

Each State may require aircraft entering its territory to land at a customs airport for a customs examination, unless the flight has permission to cross the territory without landing. Similarly, aircraft departing a State may be required to depart from a customs airport.

Article 11 – Applicability of air regulations

Any aircraft, regardless of its nationality, shall obey the regulations and operational procedures of the State in which it is flying.

Article 12 – Rules of the Air

Each State shall ensure that aircraft operating within its territory, or aircraft carrying its nationality mark, wherever they may be, follow the rules of the air. Over the high seas, the rules of the Convention apply. Each State shall endeavour to prosecute violators of the regulations.

Article 13 – Entry and clearance regulations

Regulations of a State relating to entry, clearance, immigration, passports, customs and quarantine must be complied with by or on behalf of passengers, crew or cargo on entry into, departure from or while within the territory of that State.

Article 16 – Search of aircraft

Each State has the right to search aircraft from other States on landing or prior to departure, and to inspect documents.

Article 17 – Nationality of aircraft

Aircraft have the nationality of the State in which they are registered.

Article 18 – Dual registration

An aircraft may not be registered in more than one State, though its registration may be changed from one State to another.

Article 19 – National laws governing registration

Registration or transfer of registration in any State shall comply with that State's laws and regulations.

Article 20 – Display of marks

All aircraft operating internationally shall display their appropriate nationality and registration marks.

Article 22 – Facilitation of formalities

Each State shall facilitate flights between territories of contracting States, and prevent unnecessary delays to those flights, especially in relation to customs, immigration and quarantine procedures.

Article 23 – Customs and immigration procedures

Each State shall establish customs and immigration procedures in accordance with international practice.

Article 24 – Customs duty

Aircraft entering another State's territory shall be admitted temporarily free of duty, subject to the State's customs regulations. Fuel, oil, spare parts and regular equipment that are on board an aircraft on arrival in another State, and retained on board on departure, shall be exempt from duty. This does not apply to anything that is unloaded from the aircraft.

Spare parts imported into a State for use by an aircraft from another State on international operations shall be free of duty.

Article 25 – Aircraft in distress

Each State shall assist aircraft in distress in its territory, and allow the owners of the aircraft and that State in which the aircraft is registered to assist as appropriate.

Article 26 – Investigation of accidents

Should an aircraft registered in one State be involved in an accident in another State, and the accident results in death or serious injury or indicates a serious technical defect in the aircraft or navigation facilities, the State in which the accident occurs shall carry out an inquiry in accordance with ICAO procedures. The State in which the aircraft is registered shall be allowed to observe the inquiry.

Article 28 – Air navigation facilities and standard systems

Each State shall facilitate international aviation by:
- providing radio services, meteorological services and air navigation facilities to ICAO standards;
- operating standard systems for communications, markings, signals and lighting;
- cooperating internationally in the publication of aeronautical maps and charts.

Article 29 – Documents carried in aircraft

All aircraft flying internationally shall carry the following documents:
- Certificate of Registration;
- Certificate of Airworthiness;
- appropriate licences for each crew member;
- journey logbook;

- appropriate radio licences;
- if carrying passengers, a list of their names and places of embarkation (boarding) and destination;
- if carrying cargo, a manifest and detailed declarations of cargo.

Article 30 – Aircraft radio equipment

Aircraft operating in other States may carry radio transmitting equipment only if it is licensed by the State in which the aircraft is registered. The use of that equipment shall comply with the regulations of the State that is being flown over.

Radio transmitting equipment may be used only by crew members who are licensed to do so by the State in which the aircraft is registered.

Article 31 – Certificates of airworthiness

All aircraft operating internationally must have a valid Certificate of Airworthiness issued by the State in which it is registered.

Article 32 – Licences of personnel

Pilots and flightcrew members engaged in international operations shall hold licences issued by the State in which the aircraft is registered.

For flight over its own territory, each State reserves the right to refuse to recognise flight crew licences and certificates of competency granted to its nationals by another State.

Article 33 – Recognition of certificates and licences

Certificates of airworthiness and flightcrew licences issued by the State in which the aircraft is registered shall be recognised by other contracting States, provided the requirements for the issue of such certificates and licences meet ICAO standards.

Article 34 – Journey logbooks

All aircraft on international operations shall keep a journey logbook, containing details of the aircraft, its crew and each journey, in accordance with ICAO standards.

Article 35 – Cargo restrictions

Weapons or munitions of war may not be carried in or above a State's territory except by permission of that State.

States may prohibit the carriage of any other items within their territory for reasons of public order or safety.

Article 36 – Photographic apparatus

States may prohibit or regulate the use of photographic apparatus in aircraft over its territory.

Article 37 – Adoption of international standards and procedures

Each State undertakes, as far as possible, to implement uniformity in aviation regulations, standards and procedures. To help this process

ICAO shall adopt and amend international standards and recommended practices dealing with such matters as:

- communications systems and air navigation aids, including ground markings;
- airports and landing areas;
- rules of the air and air traffic control procedures;
- licensing of flightcrew and maintenance staff;
- airworthiness of aircraft;
- meteorological services;
- logbooks;
- aeronautical maps and charts;
- customs and immigration procedures;
- aircraft in distress and accident investigation.

Article 39 – Endorsement of certificates and licences

Aircraft that have failed to meet any international standard of airworthiness or performance at certification shall show on its airworthiness certificate full details of such failure(s).

Flightcrew licence holders who fail to satisfy any condition laid down in the international standard relating to that licence shall have full details of such failure(s) shown on their licence.

Article 40 – Validity of endorsed certificates and licences

Aircraft and flightcrew may operate internationally only if their certificates or licences permit it. The use of any aircraft or certificated aircraft part in a State other than the one in which it was originally registered shall be at the discretion of the State into which the aircraft or part is imported.

Article 43 – ICAO name and composition

An organisation named The International Civil Aviation Organization is formed by the Convention. It is made up of an Assembly, a Council, and other necessary bodies.

Article 44 – ICAO objectives

The Organization aims to develop international air navigation and international air transport so as to:

- ensure the safe and orderly growth of civil aviation throughout the world;
- encourage aircraft design and operation for peaceful purposes;
- encourage development of airways, airports and navigation facilities for international civil aviation;
- provide the world with safe, regular, efficient and economical air transport;
- promote fair competition and avoid discrimination between States;
- promote flight safety.

Article 47 – Legal capacity

Each State shall grant to ICAO such legal capacity as may be necessary to perform its functions.

ICAO Annexes

The terms used in ICAO Annexes are summarised on page 173. You should become familiar with these terms.

The ICAO Articles are augmented by numerous international conferences and committees to discuss and agree various aviation issues such as: Rules of the Air, Operation of Aircraft, Aerodromes, Air Traffic Services etc. This information is disseminated in the form of 18 **Annexes**. These are supplementary documents detailing information on specific subjects and completely define aviation terms, standards and recommended practices. These Annexes are regularly reviewed and amended and may themselves be further supplemented by regional procedures where necessary.

Extracts from some of these Annexes applicable to student and private pilots are set out in later chapters, with the UK differences where applicable. The paragraph numbers refer to the paragraph numbers in the Annex concerned, should you wish to do further research.

NOTE National differences from the ICAO system that prevail in the UK are shown in AIP GEN 1-7. You need to be familiar with these national differences as well as with the international system, since these form part of the PPL examinations.

Joint Aviation Authority (JAA)

Of the 37 States within the JAA, the following 26 are full members: Austria, Belgium, Czech Rep., Denmark, Finland, France, Germany, Greece, Hungary, Iceland, Ireland, Italy, Luxembourg, Malta, Monaco, Netherlands, Norway, Poland, Portugal, Romania, Slovenia, Spain, Sweden, Switzerland, Turkey, United Kingdom.

The United Kingdom is an active member of ICAO, as are most large nations. While standardisation has occurred to a large extent, national differences remain. European nations, though close to each other geographically, ended up with many variations in their aviation systems and regulations.

Since the foundation of the European Union (EU), successive British governments have sought to standardise aviation practices with our European partners. This led to the harmonisation of flight crew licensing and a code of practice was adopted by EU member States.

An organisation called the Joint Aviation Authority (JAA) was formed. At present 37 States belong to the JAA, of which 26 are full member States (including the UK), and 11 are candidate members who have agreed to comply with the organisation's requirements and attain full membership in due course. No doubt more States will follow. However, the JAA has no legal status, because the member States have not yet entered into any form of treaty. Since some of the JAA States are not members of the EU, European legislation cannot cover JAA either.

Moves are afoot to end this unsatisfactory state of affairs by the introduction of a new body known as the European Aviation Safety Agency (EASA). EASA has far reaching regulatory powers and in the future this will particularly affect pilot licensing. This will have a significant effect on all levels of aviation (except for Very Light Flying Machines for recreational use).

The main regulatory setting up of the system was adopted in September 2003, and its main impact will be felt by the aviation industry within the next 3–5 years. The Basic Regulation only deals with certification of aeronautical products, parts and appliances and the approval of organisations and personnel employed in the maintenance of such products. The European Union will propose legislation to extend its scope in the coming year.

A transitional period of 5 years has been allowed whereby member States can continue to issue certificates until the Agency is ready to take over.

More detail on JAA and JAR-FCL, which covers the rules on flight crew licensing, can be found in Chapter 10.

Aviation Law in the United Kingdom

Aviation law in the United Kingdom is enacted by Parliament and published in statutory documents. The principal source of regulations for the private pilot is the **Air Navigation Order** (see page 14). Another is the **Air Navigation (General) Regulations.**

The authority responsible for civil aviation in the United Kingdom is the **Civil Aviation Authority** (CAA). Article 129 defines the 'competent authority' as meaning in relation to the UK, the CAA. (In relation to any other country, it means the authority responsible under the law of that country for promoting the safety of civil aviation.) One of its main functions is to provide an **Aeronautical Information Service** (AIS) to collect and disseminate the information necessary for safe and efficient air navigation. It does this through four documentation channels:

1. On 1 January 1998, the United Kingdom **Aeronautical Information Publication** (UK AIP) was significantly revised to comply more closely with European requirements. It is now published in three parts, with a regular amendments service, and contains information under the following headings:

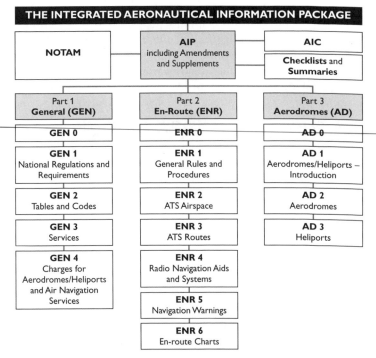

■ *Figure 1-1* **The UK AIP and related documents**

NOTE ICAO Annex 15 deals with international requirements.

2. **NOTAMs** (Notices to Airmen), distributed by the aeronautical teleprinter network, contain information on any aeronautical facility, service, procedure or hazard – timely knowledge of which is required by people concerned with flight operations and which may be given one of three categories:

NOTAM CATEGORIES	
NOTAMN	*new information*
NOTAMR	*replacing previous information*
NOTAMC	*cancelling a previous NOTAM*

3. **Pre-flight Route and Aerodrome Information Bulletins** (GENs) should be readily available to you. They are a vital part of your pre-flight planning. Daily updated information and answers to queries may be obtained from the Aeronautical Information Service (AIS), tel: 0208 754 3464, or on the Internet at www.ais.org.uk, together with the NOTAMS issued as the UK Preflight Information Bulletins. Freephone updated information on Royal Flights and Temporary Restricted Airspace for the *Red Arrows* display team is available daily on 0500 354802.

Aeronautical Information Circulars (AICs) are published monthly and concern administrative matters and advance warnings of operational changes, and draw attention to and advise on matters of operational importance, e.g. the availability of aeronautical charts, corrections to these charts and amendments to the Chart of UK Airspace Restrictions. AICs are colour-coded according to their subject matter:

AERONAUTICAL INFORMATION CIRCULARS (AICs)	
Subject	Colour of paper
Air Safety	pink
Administrative	white
Operational and Air Traffic Services	yellow
UK Restrictions Charts	mauve
Maps/Charts	green

NOTE that the *Aviation Law, Flight Rules and Operational Procedures* examination may refer to the contents of a specific AIC. You should read all relevant AICs, particularly the more important pink ones, which will be available in a folder at your flying training organisation or on the Internet at www.ais.org.uk. These include:

IMPORTANT AIR SAFETY AICs	
AIRPROX Reporting	AIC 15/1999 (Pink 186)
Danger Areas – Advice to Pilots	AIC 87/2002 (Pink 39)
Frost, Ice and Snow on Aircraft	AIC 93/2000 (Pink 8)
Icing (Induction System) on Piston Engines	AIC 145/1997 (Pink 161)
Low-Altitude Windshear	AIC 19/2002 (Pink 28)
Medication, Alcohol and Flying	AIC 58/2000 (Pink 4)
Take-Off, Climb and Landing Performance	AIC 67/2002 (Pink 36)
Effect of Thunderstorms and Associated Turbulence	AIC 72/2001 (Pink 22)
Wake Turbulence	AIC 17/1999 (Pink 188)
Use of Portable Telephones in Aircraft	AIC 62/1999 (Pink 196)

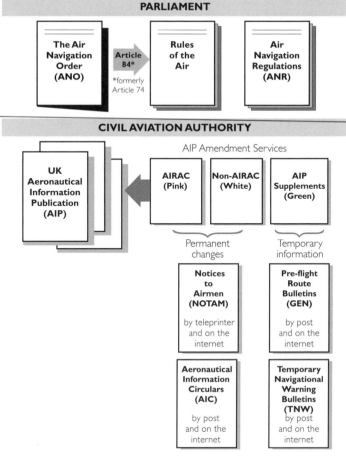

■ *Figure 1-2* **The regulatory documents**

Explanatory Publications

Since all the documents detailed must cover the full range of aviation activity, their contents in total are quite daunting for a private pilot. To present vital matters in a simpler form, the CAA produces subsidiary Civil Aviation Publications (CAP), including:

- **LASORS** – Licensing, Administration, Standardisation, Operating Requirements, Safety – probably the best source of reference and guidance material produced by the CAA.
- **CAP 413** on Radiotelephony (R/T or RTF) procedures;
- **CAP 637**, an excellent guide to visual aids and signals for pilots and ground personnel.

The CAA also publishes an invaluable series of *General Aviation Safety Sense* leaflets (also incorporated into LASORS), which are generally available at flying schools and aero clubs, or can be obtained free of charge from:
  Documedia Ltd, 37 Windsor Street, Cheltenham, Glos., GL52 2DG. Tel: (01242) 235151. See also CAA website.

There are also some non-CAA documents that are useful in clarifying operational practices, such as *Pooley's Flight Guide*, published annually and available throughout the UK.

The Air Navigation Order 2000 (ANO)

The Air Navigation Order (also referred to as the 'Order' or as the 'ANO') is a *statutory instrument* that was enacted by Parliament to form the basis of civil aviation in the UK.

Parts and Articles in the ANO
The Air Navigation Order is arranged in **Parts**, each relating to an item of major importance, e.g. Part (viii) – *Movement of Aircraft.*

The Parts of the ANO each contain a number of **Articles** and these are numbered consecutively from the start of the ANO, dealing with particular subjects within the broader scope of the Part. For example, Part (viii) commences with Article 84, *Rules of the Air and Air Traffic Control.*

Although the rules themselves are listed in a separate statutory instrument with the title *The Rules of the Air and Air Traffic Control Regulations,* their basis is Article 84 (previously 74) of the ANO.

Appendices in the ANO
Further detail in the ANO is given in a series of Appendices to the Order, known as **Schedules**, e.g. *Schedule 4, Aircraft Equipment.* Operators would refer to this to determine the minimum equipment necessary when fitting out aircraft for a particular task.

People concerned with planning and developing flight operations will constantly refer to the ANO, e.g. to determine requirements in such matters as equipment, licensing of crew members, documents and records, registration of aircraft and their operation. The private pilot, however, will probably find that only occasional reference to specific Articles in the ANO is sufficient, although some knowledge of the contents is required.

General

Penalties (ANO Article 122)

Pilots who contravene the Air Navigation Order (ANO) or Regulations made under the Order are held liable, and this Article defines the extent of liability and the maximum penalties which may be imposed if convicted of such contraventions.

Interpretation (ANO Article 129)

Article 129 gives the legal definitions for the terms used in the ANO, many of which concern the private pilot and some of which are commonly misconstrued. It also acts as a glossary to the ANO.

Now complete **Exercises 1 – Aviation Law and Legislation**

Rules of the Air

Rules of the Air (UK ANO)

The Rules of the Air have been established to allow aircraft to operate as safely as is reasonably possible. Air safety depends to a large extent on all pilots understanding the basic rules and operating within them.

ICAO Annex 2 sets out the international standards of the Rules of the Air. The UK and its European partners in the JAA fully endorse these international standards. The UK has enacted the regulations under Article 84 of the ANO (previously Article 74).

The rules are known by their numbers; Rule 5, for example, concerns low flying. They are referred to, where applicable, throughout this section on aviation law. In this chapter we will be dealing with those topics covered in the rules which are applicable to aerial operations by a Private Pilot's Licence holder.

The Rules of the Air apply to:
- all aircraft in the UK (including the neighbourhood of offshore installations, where the low-flying rule is concerned); and
- to all UK-registered aircraft wherever they may be.

The rules may be departed from to the extent necessary for avoiding immediate danger or for complying with the law of any other country in which the aircraft might be. When a rule is departed from for safety reasons, the circumstances must be reported afterwards to the competent authority.

Collision Avoidance in the Air (Rule 17)

With many aircraft sharing the same airspace, it is often necessary to take collision avoidance action. A collision risk exists when one aircraft is at the same level or approaching another, its range is decreasing and its relative bearing remains constant (Figure 2-1).

Some basic rules understood by all pilots, and applied when necessary, are essential to avoid aerial collisions.

General

- Regardless of any ATC clearance, it is the duty of the *commander* (pilot-in-command) of an aircraft to take all possible measures to see that he does not collide with any other aircraft.
- An aircraft must not fly so close to other aircraft as to create a danger of collision.
- Aircraft must not fly in formation unless the commanders have agreed to do so.

- ☐ An aircraft which is obliged to give way to another aircraft must avoid passing over, or under, or crossing ahead of, the other aircraft (unless passing well clear of it).
- ☐ An aircraft with right of way should maintain its heading and speed.
- ☐ For the purposes of this rule, a glider and a machine towing it are considered to be a single aircraft under the command of the commander of the towing machine. Aeroplanes and helicopters must give way to aircraft towing gliders.

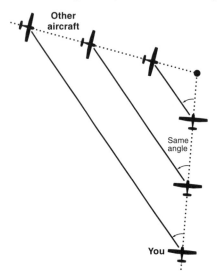

■ *Figure 2-1* **A constant relative bearing**

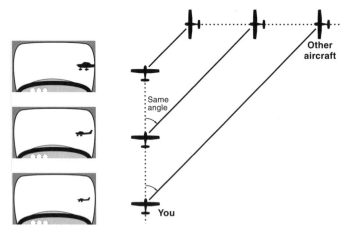

■ *Figure 2-2* **Fixed position of another aircraft in the windscreen indicates a constant relative bearing and therefore a collision risk**

Approaching Head-On

When two aircraft are approaching head-on or nearly so, and there is danger of collision, each must turn right (Figure 2-3).

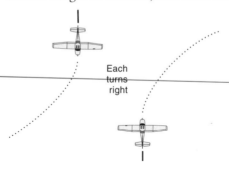

Each
turns
right

■ *Figure 2-3* **Approaching head-on – each turns right to avoid a collision**

Overtaking

An aircraft which is being overtaken in the air has right of way and the overtaking aircraft, whether climbing, descending or level, must keep out of the way by turning right (Figure 2-4). An overtaking situation is considered to be that where the overtaking aircraft is within 70° of the overtaken aircraft's centreline.

A glider overtaking another glider may, however, turn either right or left.

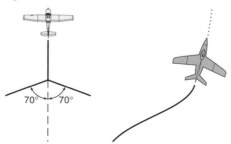

70°   70°

■ *Figure 2-4* **Overtaking – keep right**

Converging Aircraft

▢ An aircraft in the air must give way to other converging aircraft as follows:
- flying machines must give way to airships, gliders and balloons;
- airships must give way to gliders and balloons;
- gliders must give way to balloons.

▢ Subject to the above paragraph, when two aircraft are converging at about the same altitude, the aircraft which has the other on its right must give way. Powered aircraft must give way, however, to those towing other aircraft or objects such as banners.

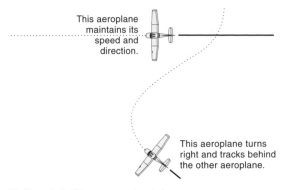

This aeroplane
maintains its
speed and
direction.

This aeroplane turns
right and tracks behind
the other aeroplane.

■ *Figure 2-5* **Give way to the right**

Flight in the Vicinity of an Aerodrome (Rules 17 & 35)

A flying machine, glider or airship flying in the vicinity of an aerodrome or moving on an aerodrome shall, unless the aerodrome's ATC unit otherwise authorises:

☐ conform to the pattern of traffic formed by other aircraft intending to land at that aerodrome, or keep clear of the airspace in which the pattern is formed;

☐ make all turns to the left unless ground signals direct otherwise.

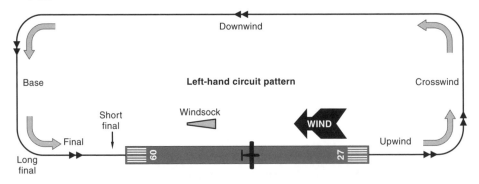

■ *Figure 2-6* **Make left-hand circuits, unless right-hand circuits are indicated**

NOTE Sometimes ATC approves a straight-in approach to an arriving aircraft, which will then normally report "long final" at 8 nm, and "final" at 4 nm from the runway.

Order of Landing

An aircraft landing or on final approach has right of way over others in flight or on the surface.

In the case of two or more flying machines, gliders or airships approaching any place for landing, the lower aircraft has right of way (although it must not cut in front of, or overtake, another which is on final approach); provided that:
- when ATC has given any aircraft an order of priority for landing they must approach to land in that order; and
- when the commander of an aircraft is aware that another is making an emergency landing he must give way, and at night (even if he already has permission to land) must not attempt to land until given further permission.

Landing and Take-Off

A flying machine, glider or airship must take off and land in the direction indicated by ground signals or, in their absence, into wind unless good aviation practice demands otherwise.

A flying machine or glider must not land on a runway which is not clear of other aircraft, unless an aerodrome ATC unit otherwise authorises. If an aircraft is on the runway, Air Traffic Control may give you an instruction "land after", which is **not** a clearance to land. The decision is yours, based on whether you judge that sufficient separation exists. Such an instruction is never issued at night.

Where take-offs and landings are not confined to a runway:
- a flying machine or glider when landing must leave clear on its left any aircraft which has landed, or is already landing, or is about to take off (i.e. keep to the right of other aircraft). If such a flying machine or glider is obliged to turn when taxiing on the landing area, it shall turn to the left after the commander has satisfied himself that such action will not interfere with other traffic movements; and
- a flying machine about to take off must manoeuvre so as to leave clear on its left any aircraft which has taken off, or is about to take off.

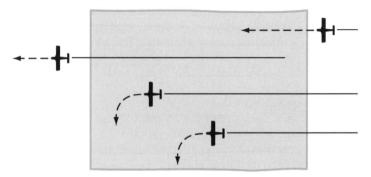

■ *Figure 2-7* **Turn left after landing when operations are not confined to a runway**

▢ A flying machine after landing must move clear of the landing area as soon as possible unless the aerodrome ATC unit otherwise authorises.

Use Radio in an Aerodrome Traffic Zone (Rule 39)

When flying in an Aerodrome Traffic Zone, a pilot must maintain appropriate radio (or other) communication with the aerodrome authority (Air Traffic Control, Aerodrome Flight Information Service or Air/Ground radio).

NOTE Aerodrome Traffic Zones (ATZs) are covered in detail in Chapter 5, *Airspace*.

Right-Hand Traffic Rule (Rule 19)

An aircraft flying in sight of the ground and following a road, railway, canal, coast or other line feature shall keep the line feature on its left, except when the aircraft is flying in controlled airspace, and where it is instructed to do otherwise by the appropriate Air Traffic Control authority. This will ensure separation from aircraft flying in the opposite direction and following the same line feature.

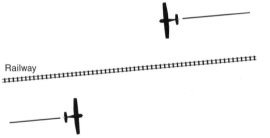

■ *Figure 2-8* **Keep to the right (and keep the line feature on your left)**

Lights on Aircraft (Rules 8–15)

The direction of flight of aircraft is more difficult to determine by night than by day. To assist in the identification of aircraft position and heading by night, aircraft must display such lights as are specified for the particular category of aircraft. These lighting requirements generally apply when the aircraft is moving on the ground also. No other lights may be displayed that would impair the effectiveness of the required lights.

If any required light fails in flight and cannot be repaired or replaced at once the aircraft must **land** as soon as it can safely do so unless ATC authorises continuation of the flight.

Knowing the arcs of the basic aircraft navigation lights helps in assessing collision risk. On flying machines and airships the **green** main navigation light on the **right** (or starboard) wing and **red** main navigation light on the **left** (or port) wing show through 110 degrees from dead ahead out to their respective sides. The white tail-light shows through 70 degrees either side of dead astern. An anti-collision light, where carried on an aircraft, is a flashing red light showing in all directions.

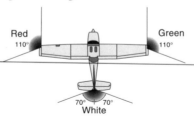

■ *Figure 2-9* **Aeroplane lights**

If at night you see the red navigation light of an approaching aircraft out to the left, then your paths will not cross (i.e. red to red is safe). A red light out to the right, however, could mean that the risk of a collision exists.

The situation is reversed for the green light of an approaching aircraft – out to the right it is safe, out to the left there is a risk of collision. Green to green is safe.

If both the red and green lights are visible, then the other aeroplane is flying directly towards you. If the white light is visible, it is flying away from you.

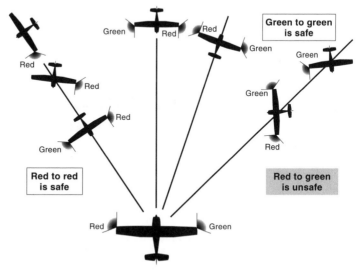

■ *Figure 2-10* **Using navigation lights to avoid collision**

In the case of airships there is also a white nose-light showing through 110 degrees either side of dead ahead. A glider may show either the basic lights for a flying machine, as just described, or a steady red light visible in all directions. Free balloons are required to show a steady red light visible in all directions.

In the United Kingdom, **night** is defined for the Rules of the Air as being:
- from 30 minutes after sunset;
- until 30 minutes before sunrise.

Low Flying Regulations (Rule 5)

The following restrictions apply to low flying in all cases other than the exceptions specified on page 26.

Flight over Congested Areas

A **congested area** in relation to a city, town or settlement, means any area which is substantially used for residential, industrial, commercial or recreational purposes.

An aircraft (other than a helicopter) must not fly over a congested area:
- below a height that would allow it to **land clear** of the area and without danger to people or property if an engine fails; or
- less than **1,500 feet above** the highest fixed object within 600 metres of the aircraft, whichever is the higher.

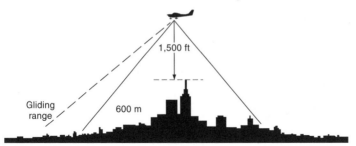

■ *Figure 2-11* **'Land clear' and 1,500 ft clearance over congested areas**

If the aircraft is towing a banner, the height must be calculated on the basis that the banner is not dropped within the congested area if an engine fails.

Large Open-Air Gatherings

No aircraft may fly over or within **1,000 metres** of an open-air gathering of more than 1,000 people except with the permission in writing of the Authority (the CAA) and in accordance with any conditions specified, and with the written consent of the organisers; nor may it fly below a height that would enable it to **land clear** of the assembly if an engine failed.

If towing a banner, the banner shall not be dropped within 1,000 metres of the assembly.

(In the case of inadvertent contravention of this rule, it may be a good defence to show that the particular flight was made at a reasonable height and for a purpose not connected with the assembly.)

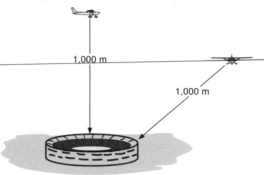

■ Figure 2-12 **Maintain 1,000 metres clearance from large open-air gatherings**

The 500 ft Rule

An aircraft must not fly closer than 500 feet to any person, vessel, vehicle or structure. (Exceptions to this rule are included below.)

■ Figure 2-13 **Maintain 500 ft clearance**

Helicopters

A helicopter must not fly below a height that would allow it to land without danger to people or property if an engine fails. Nor (except with the written permission of the Authority) may it fly over a congested area at less than 1,500 feet above the highest fixed object within 600 metres, or over a specified area of Central London below a height that would enable it to land clear of that area if an engine fails.

NOTE The general requirement for all aircraft to be able to land clear is the responsibility of the pilot-in-command, and not of ATC. If an ATC clearance is given which would not permit the requirement to be met, e.g. a Special VFR clearance at 1,000 feet, say, in a single-engined aircraft over London, the clearance should not be accepted by the pilot-in-command.

Exceptions

☐ The proviso stipulating '1,500 feet above the highest fixed object within 600 metres' does not apply on routes notified for the purposes of the Low Flying Regulations (Rule 5), or on Special VFR flights made in accordance with ATC instructions.

☐ The '1,000 metres' and '500 feet' rules specified above do not apply to flights over or within 1,000 metres of an assembly gathered for an aircraft race, contest or display if the aircraft is part of the event and the pilot-in-command has the written consent of the organisers.

☐ The '500 feet' rule specified above does not apply to any aircraft landing or taking off in accordance with normal aviation practice, nor to a glider which is hill-soaring. However, when practising engine failure after take-off drills, it is still the responsibility of the pilot-in-command to comply with the '500 feet' rule.

☐ None of the above rules apply when flying as necessary to save life or during take-off, landing, practice approaches and navigation aid checking at a Government, Authority or licensed aerodrome. Practice approaches must, however, be made in airspace customarily used for landing or taking off. Note that the '1,500 feet' prohibition specified above *does* apply to flight after take-off or before landing at *unlicensed* aerodromes, i.e. no circuit practice is permitted at unlicensed aerodromes.

WARNING Be aware that the CAA will prosecute in cases of alleged low flying. Not only is it illegal to carry out low-level flypasts, it does nothing for the image of light aviation in the eyes of those people who campaign regularly for the closure of airfields. Even if the prosecution is not proved, you will have the expense of defending the case. If you lose the case, you will have the stigma of a criminal record, and, possibly, incur a heavy fine and the CAA's legal costs. In addition, low-level flying is simply dangerous. High-tension cables, aerials and masts may not be marked on charts and are often difficult to see. Leave low flying to the military and crop sprayers, who have been trained properly for this role.

Reporting Hazardous Conditions (Rule 4)

The commander of an aircraft must report as soon as possible (to ATC) any hazardous flight conditions encountered, giving details pertinent to the safety of other aircraft. Typical situations worthy of reporting are severe windshear or turbulence, rapidly deteriorating visibility or an unserviceable runway.

Aerobatics (Rule 18)

Aerobatics are not permitted over a congested area. (A 'congested area' is one which is substantially used for residential, industrial, commercial or recreational purposes within a city, town or settlement.) Within controlled airspace aerobatics may be permitted with the specific approval of the controlling authority.

Simulated Instrument Flight (Rule 6)

When simulated instrument flying is taking place (i.e. a pilot is flying the aircraft with his external field of view artificially restricted), the aircraft must have dual flying controls, and a second pilot (known as a *safety pilot)* must be present who can assist the other (note that a student pilot does not qualify for this assistant role). If necessary, a third person is to be carried as an observer to ensure an adequate lookout.

Practice Instrument Approaches (Rule 7)

A pilot practising instrument approaches in Visual Meteorological Conditions (VMC) must inform ATC and carry a 'competent observer'.

Misuse of Signals and Markings (Rule 3)

Signals and markings specified in the rules for a particular meaning or purpose must not be used except with that meaning or for that purpose. Signals which may be confused with a specified signal must not be made, and military signals must not be used except with lawful authority.

Rules of the Air (ICAO Annex 2)

In addition to the Rules of the Air, pilots making international flights need to be familiar with interception procedures.

Interception of Civil Aircraft

In order to comply with ICAO standards, pilots leaving UK airspace to make international air navigation flights are required to carry a copy of the procedures to be followed in the event of interception of their aircraft for reasons of military necessity, public safety or to prohibit flight over restricted or danger areas of another State's territory. You will find these procedures in the CAA's *General Aviation Safety Sense Leaflet no. 11, Interception Procedures.* A summary of this leaflet follows.

Interception is rare, but if you are intercepted the most important things to do are to keep calm, and to follow precisely any instructions you are given.

Procedures

If you are intercepted by another aircraft you must immediately:

- [ ] follow the instructions given by the intercepting aircraft, interpreting and responding to visual signals as described below;
- [ ] notify, if possible, the appropriate Air Traffic Services Unit;
- [ ] attempt to establish radio communication with the intercepting aircraft or control unit on 121.50 MHz, giving your callsign and the nature of your flight;
- [ ] squawk code 7700 on your transponder, unless otherwise instructed by the ATSU.

Signals from intercepting aircraft

The standard signals used by intercepting aircraft and the responses you (the intercepted aircraft) should give are summarised below.

INTERCEPTING AIRCRAFT	INTERCEPTED AIRCRAFT
"You have been intercepted. Follow me."	**"Understood, will comply."**
The intercepting aircraft rocks its wings from a position slightly above and ahead of your aircraft. After your acknowledgement, it makes a slow level turn, normally to the left, onto the desired heading.	*Rock your wings and follow the intercepting aircraft* **immediately**.

NOTE At night, flash your navigation lights at irregular intervals in addition to the above response. If your aircraft cannot fly fast enough to keep up with the intercepting aircraft, it will fly a series of racetrack patterns and rock its wings each time it passes you.

The signal to indicate that you may proceed with your flight is:

INTERCEPTING AIRCRAFT	INTERCEPTED AIRCRAFT
"You may proceed."	**"Understood, will comply."**
The intercepting aircraft makes an abrupt break-away manoeuvre from your aircraft, consisting of a climbing turn of 90° or more without crossing the line of flight of your aircraft.	*Rock your wings.*

If the intercepting aircraft wants you to land, it will signal as below:

INTERCEPTING AIRCRAFT	INTERCEPTED AIRCRAFT
"Land at this aerodrome."	**"Understood, will comply."**
The intercepting aircraft circles the aerodrome, lowering its landing gear and overflying the runway in the direction of landing. At night the intercepting aircraft will also show steady landing lights.	*Lower landing gear (if possible), follow intercepting aircraft and, if after overflying the runway you consider landing is safe, do so.*

Signals from intercepted aircraft

Should you need to signal to the intercepting aircraft, you may make the following standard signals:

INTERCEPTED AIRCRAFT	INTERCEPTING AIRCRAFT
"Aerodrome designated is inadequate."	**"Follow me to an alternate aerodrome."**
Raise your landing gear (if possible) while overflying landing runway at a height above 1,000 ft and below 2,000 ft above the aerodrome level, and continue circling the aerodrome.	Intercepting aircraft will raise its landing gear and rock wings to indicate "Follow me." OR **"You may proceed."** Intercepting aircraft makes an abrupt break-away climbing turn.
"I cannot comply."	**"Understood."**
Switch all available lights on and off at regular intervals, but in a manner that is distinct from flashing lights.	Intercepting aircraft makes an abrupt break-away climbing turn.
"I am in distress."	**"Understood."**
Switch all available lights on and off at irregular intervals.	Intercepting aircraft makes an abrupt break-away climbing turn.

NOTE Avoid using hand signals because they could be misinterpreted.

Radio communication

If you establish radio contact with the intercepting aircraft but do not share a common language, use the following phrases to acknowledge instructions and convey essential information.

INTERCEPTED AIRCRAFT		
Phrase	**Pronunciation**	**Meaning**
Callsign	**Kol** sa-in	My callsign is (callsign)
Wilco	**Vill**-co	Understood, will comply
Can not	**Kann** nott	Unable to comply
Repeat	Ree-**peet**	Repeat your instruction
Am lost	**Am losst**	Position unknown
Mayday	**Mayday**	I am in distress
Hijack	**Hijack**	I have been hijacked
Land (place-name)	**Laand** (place name)	I request to land at (place-name)
Descend	Dee-**send**	I require descent

The following phrases should be used by the intercepting aircraft:

INTERCEPTING AIRCRAFT		
Phrase	**Pronunciation**	**Meaning**
Callsign	**Kol** sa-in	My callsign is (callsign)
Follow	**Fol**-lo	Follow me
Descend	Dee-**send**	I require descent
You land	**You laand**	Land at this aerodrome
Proceed	Pro-**seed**	You may proceed

If you receive instructions from another source that conflict with those given by the intercepting aircraft's visual signals or radio instructions, you shall request immediate clarification while continuing to obey the visual instructions. Again, it is not a good idea to try and make hand signals, they can be easily misunderstood – not a good situation if you are head-to-head with a jet fighter!

Finally, if you are ever intercepted in foreign airspace, tell the CAA's Safety Data & Analysis Unit about it. As interceptions are rare, your experience may provide useful information to others.

Signals from the ground
Visual signals may be used to warn unauthorised aircraft flying in, or about to enter a restricted, prohibited or danger area. These are, by day or night, a series of projectiles discharged from the ground at intervals of 10 seconds, each showing, on bursting, red and green stars or lights. This will indicate to an unauthorised aircraft that it is flying in, or about to enter a restricted, prohibited or danger area, and that the aircraft is to take such remedial actions as may be necessary (normally to leave the area as quickly as possible without descending).

Now complete **Exercises 2 – Rules of the Air.**

Aerodromes

Aerodromes (UK ANO)

General Characteristics

An aerodrome is an area of land or water used for the taking off and landing of aircraft. Aerodromes are divided into various categories depending on their use. The *Aerodromes* (AD) section of the UK Aeronautical Information Publication (AIP) contains an aerodrome directory giving specific information on physical characteristics, local hazards and flying restrictions.

Limitations on the Use of Aerodromes

Certain restrictions apply at some aerodromes and not at others.

At military aerodromes, and at civil aerodromes with an ordinary licence, **prior permission to land** is needed from the aerodrome authority, and at unlicensed aerodromes the **prior permission of the owner or person in charge**. This may be designated in documents as 'PPR', which stands for *prior permission required*.

PPR usually means that you should telephone the airfield operator or owner before departure, particularly if there is no radio frequency published for the airfield. Remember that if you simply turn up and land without seeking permission you are trespassing on private property. Further, there may be safety implications as a particular airstrip may be usable only in a certain direction or in certain wind conditions. It may be prone to waterlogging or there could be specific noise abatement procedures to be followed.

Failure to seek permission and the requisite information could lead to an accident, and subsequent refusal by the insurance company to pay for the claim because you would have breached the Air Navigation Order.

Permission to use a military aerodrome must always be obtained before taking off for the aerodrome concerned. It is likely that you will have to demonstrate a higher level of insurance cover to land at a military aerodrome.

The civil use of military aerodromes is restricted to the normal hours of watch, and to aircraft on inland flights. At some military aerodromes, civil use is further restricted to certain classes of traffic (e.g. scheduled services, charter flights or private aircraft).

The above restrictions on aerodrome use would not apply in the case of an in-flight emergency. Aerodromes not listed in AIP AD (*Aerodromes* section of the Aeronautical Information Publication) may be used in an **emergency** or if **prior permission** from the owner or operator is obtained.

Rules of the Air relating to Aerodromes

Use of Aerodromes for Instruction in Flying (ANO Article 101)

Aeroplanes and **rotorcraft** are restricted to take off and land only at a licensed aerodrome, a Government aerodrome, or an aerodrome owned or managed by the Authority, whenever engaged in both flight instruction and flight tests for the purpose of qualifying for a pilot's licence, aircraft rating or night rating (qualification). The exception to this is training and testing for the IMC Rating, or for aerobatic training.

Gliders are similarly restricted if engaged in flying instruction, except where they are flown under club arrangements. **Self-launching motor gliders** are exempted from the restriction when being used for gliding instruction or pilot testing, if flown under club arrangements. **Microlight aeroplanes** are exempted from the restriction when flown under club arrangements.

Aviation Fuel at Aerodromes (ANO Article 112)

No person shall cause or permit any fuel to be used in an aircraft if he knows or has reason to believe that it is not fit for such use. Fuel must not be used in aircraft unless it has been dealt with in accordance with certain stipulations for the storage and sampling of aviation fuel stocks at aerodromes.

NOTE As a guide to pilots and others, and in an attempt to avoid refuelling with the wrong type of fuel, **Avgas** equipment at aerodromes is usually marked in **red**.

Notification of Arrival and Departure (Rule 20)

If an aircraft is expected at an aerodrome the commander must inform the authorities at the aerodrome as soon as possible if the destination is changed or arrival will be delayed by 45 minutes or more. This is to avoid any unnecessary *overdue action*.

Wherever possible an aircraft's commander must report upon arrival, and prior to departure, to the appropriate authority at an aerodrome.

Customs Facilities

Designated Customs and Excise Airports for the purposes of international travel are listed in the GEN (General) section of the UK AIP, together with hours of attendance and special requirements.

Aeronautical Light Beacons

Aeronautical light beacons are installed at various civil and military aerodromes in the UK. Their hours of operation vary, but broadly speaking they can be expected to be on at night and by day in bad visibility, whenever the aerodrome is operating.

Aeronautical light beacons include:

- **Identification beacons**, which flash a two–letter Morse group every 12 seconds (**green** at civil aerodromes and **red** at military aerodromes); and

- **Aerodrome beacons** which give an alternating-colour flash signal instead (usually **white/white** or (less commonly) **white/green**). They are not normally provided in addition to an identification beacon.

AERONAUTICAL LIGHT BEACONS

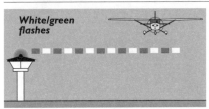

White/white flashes

Civil aerodrome beacon

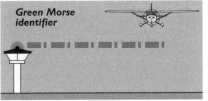

White/green flashes

Civil aerodrome beacon

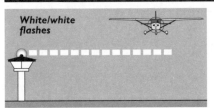

Green Morse identifier

Civil identification beacon (two-letter Morse code group, flashed every 12 seconds)

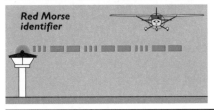

Red Morse identifier

Military identification beacon (two-letter Morse code group, flashed every 12 seconds)

Movement of Aircraft on Aerodromes (Rule 35)

An aircraft must not taxi on the apron or the manoeuvring area of an aerodrome without the permission of the aerodrome authorities.

The **manoeuvring area** of an aerodrome is that part of the aerodrome provided for the take-off and landing of aircraft and for the movement of aircraft on the surface (i.e. taxiing), excluding aprons and maintenance areas. An **apron** is a paved area of an aerodrome used for purposes such as loading and unloading of aircraft, aircraft turn-around operations, maintenance and repair, and any other approved purpose other than flight operations.

Access on Aerodromes (Rule 36)

A person shall not, without permission, go onto a part of an aerodrome provided for the use of aircraft. This applies to any part that is not a public right of way.

Right of Way on the Ground (Rule 37)

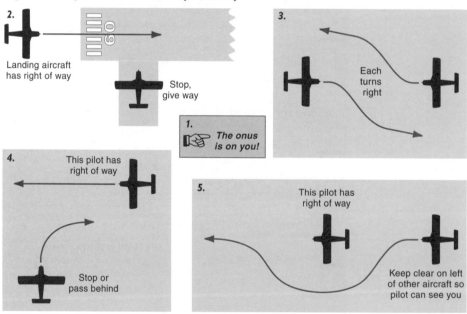

■ Figure 3-1 **Five rules for taxiing on the manoeuvring area of an aerodrome**

1. Regardless of any ATC clearance it is the duty of an aircraft commander to do all possible to avoid collision on the ground with other aircraft or vehicles.

2. Aircraft on the ground must give way to those taking off or landing, and to any vehicle towing an aircraft.

3. When two aircraft are approaching head-on, or nearly so, each must turn right.

4. When two aircraft are converging, the one which has the other on its right must give way, avoiding crossing ahead of the other unless passing well clear.

5. An aircraft which is being overtaken by another has right of way, and the overtaking aircraft must keep out of the way by turning left until past and well clear.

Aerodrome Traffic Zones (ATZs)

A certain amount of airspace surrounding most aerodromes in the UK has been designated as Aerodrome Traffic Zones, usually because of the intensity of aerial activity. See Chapter 5 for details of the dimensions of an ATZ and the airspace surrounding it.

Runway Characteristics

Declared distances at aerodromes are agreed by the relevant authority – in the UK this is the CAA, and the distances are published in the Aerodrome section of the AIP.

- **Take-off run available (TORA)**. The length of runway declared available and suitable for the ground run of an aeroplane taking off.
- **Take-off distance available (TODA)**. The length of the take-off run available plus the length of the clearway, if provided.
- **Accelerate-stop distance available (ASDA)**. The length of the take-off run available plus the length of the stopway, if provided.

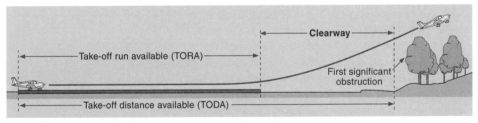

■ *Figure 3-2* **TODA, TORA and clearway**

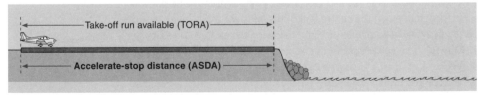

■ *Figure 3-3a* **Accelerate-stop distance (ASDA) equals TORA in this case**

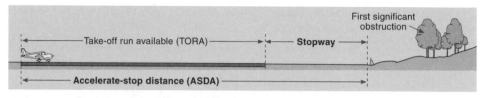

■ *Figure 3-3b* **Accelerate-stop distance (ASDA) including stopway**

■ **Landing distance available (LDA).** The length of runway declared available and suitable for the ground run of an aeroplane landing.

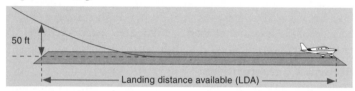

■ *Figure 3-4* **Landing distance available (LDA)**

Aerodromes (ICAO Annex 14)

Annex 14 describes ICAO standards and recommended practices for aerodromes. This Annex contains the basic specifications for the physical characteristics, configuration, performance, personnel and procedures for aerodromes that are considered desirable for safe and efficient international air navigation. Contracting States will endeavour to conform to these standards in accordance with the Chicago Convention.

Summaries of extracts relevant to PPL candidates follow. Remember that the small paragraph numbers refer to those in the ICAO Annex in case you want to study further. It is also well worth reading the CAA's *Visual Aids Handbook* (CAP 637), which is an excellent pictorial guide for pilots and personnel engaged in the handling of aircraft.

Finally, don't forget to review the terminology used in ICAO Annexes, which is summarised on page 173.

Water on a Runway

2.9.4 When water is present on a runway, a description of the runway surface conditions on the centre half of the width of the runway should be made available to pilots using the following terms:

■ **Damp** – the surface shows a change in colour due to moisture.
■ **Wet** – the surface is soaked but there is no standing water.
■ **Water patches** – significant patches of standing water are visible.
■ **Flooded** – extensive standing water is visible.

Note also the definitions of slush and snow on page 185.

Wind Direction Indicator

5.1.1 Aerodromes shall have at least one wind direction indicator. Wind direction indicators shall be located so as to be visible from aircraft in flight or on the movement area, and to be free from air disturbances by nearby objects.

Landing Direction Indicator

5.1.2 Where provided, a landing direction indicator should be located in a conspicuous place on an aerodrome.

Markings

5.2.1.4 Runway markings shall be white.

5.2.1.5 Taxiway markings and aircraft stand markings shall be yellow.

Runway Threshold and Wingbar Lights

5.3.10.9 Runway threshold and wingbar lights shall be fixed unidirectional lights showing green in the direction of approach to the runway.

5.3.11 A runway equipped with runway edge lights shall also be equipped with runway end lights (ideally at least six lights).

Runway end lights shall be placed at right angles to the runway axis as near as possible to the end of the runway (not more than 3 metres outside the end).

Runway end lights shall be fixed, showing red in the direction of the runway.

Aerodrome Signals and Markings (Rules 42–46)

A **signals area** is positioned near the control tower at aerodromes to allow messages to be passed to a pilot without the use of radio:
- **in flight,** by signals laid out on the ground; and
- **on the ground,** by signals hoisted up a mast located in the signals area.

Signals and Markings in the Signals Area

Direction of Take-Off and Landing

A **white 'T'** signifies that aeroplanes and gliders taking off or landing shall do so parallel with the shaft of the 'T' and towards the cross arm, unless otherwise authorised by the appropriate ATC unit.

A **white disc** at the head of the 'T' means that the direction of landing and the direction of take-off do not necessarily coincide. This latter situation may also be indicated by a **black ball** suspended from a mast. A rectangular **green flag** flown from a mast indicates that a right-hand circuit is in force.

■ *Figure 3-5* **Direction of take-off and landing**

Use Hard Surfaces Only

A **white dumb-bell** signifies that movements of aeroplanes and gliders on the ground shall be confined to paved, metalled or similar hard surfaces. The addition of black strips in each circular portion of the dumb-bell, at right angles to the shaft, signifies that aeroplanes and gliders taking off or landing must do so on a runway, but that movement on the ground is not confined to hard surfaces.

■ *Figure 3-6* **Use of hard surfaces signals**

Right-Hand Circuit

A red-and-yellow striped arrow bent through 90 degrees around the edge of the signals area and pointing in a clock-wise direction means that a right-hand circuit is in force.

■ *Figure 3-7* **Right-hand circuit indicator**

Where the circuit direction at an aerodrome is variable (left-hand or right-hand) a rectangular **red flag** on the signals mast indicates that a **left-hand** circuit is in operation. A rectangular **green flag** signifies that the circuit is **right-hand.**

Special precautions

A red square panel with a single yellow diagonal stripe means that the state of the manoeuvring area is poor and that pilots must exercise special care when landing.

■ *Figure 3-8* **Special precautions signal**

Landing Prohibited

A red square panel with a diagonal yellow cross signifies that the aerodrome is unsafe for the movement of aircraft and that landing is prohibited.

■ *Figure 3-9* **Landing prohibited signal**

Helicopter Operations

A **white 'H'** in the signals area means that helicopters must take off and land only within a designated area (that area itself being marked by a much larger white 'H').

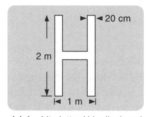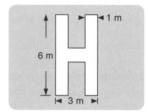

(a) A white letter H is displayed in the signals area

(b) A white letter H indicates the area to be used only by helicopters for take-off and landing

■ *Figure 3-10* **Helicopter operations markers**

Gliding

A double white cross and/or two red balls suspended from a mast, one above the other, signify that glider–flying is taking place at the aerodrome. (A similar but much larger signal is used to mark an area on the aerodrome which is to be used only by gliders).

A **yellow cross** indicates the tow-rope dropping area.

Tow-ropes, banners, etc. can only be picked up or dropped at an aerodrome, and then only as directed by the aerodrome authority, or in the designated area (yellow cross) with the aircraft flying in the direction appropriate for landing (ANO Article 55 and Rule of the Air 38).

■ *Figure 3-11* **Gliding in progress**

Signals on Paved Runways and Taxiways

Unserviceable Portion of Runway or Taxiway

Two or more white crosses along a section of runway or taxiway, with the arms of the crosses at an angle of 45 degrees to the centreline of the runway or taxiway at intervals of not more than 300 metres, signify that the section of the runway or taxiway marked by them is unfit for the movement of aircraft.

■ *Figure 3-12* **Unfit section of runway or taxiway**

Orange and white markers as illustrated in Figure 3-13, spaced not more than 15 metres apart, signify the boundary of that part of a paved runway, taxiway or apron which is unfit for the movement of aircraft. Each marker comprises a base board supporting a slatted vertical board, both of which are striped orange–white–orange.

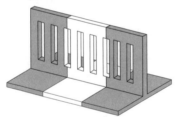

■ *Figure 3-13* **Boundary of unserviceable area marker**

Holding Point on Paved Taxiway

Parallel yellow lines – usually marked as a set of double continuous and double broken lines – across a taxiway signify a holding point, beyond which no part of an aircraft or vehicle may proceed in the direction of the runway, without ATC permission.

Of the two sets of lines, the **broken yellow lines** are located on the runway side, enabling the pilot to determine if the holding point affects him. Moving in the reverse direction towards a holding point, with the broken yellow lines encountered first (for example, having turned off the runway after landing), the holding point does not require a clearance to cross it.

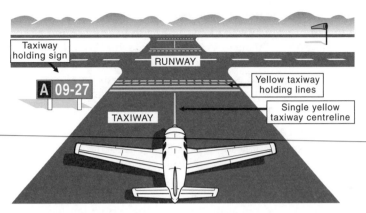

■ *Figure 3-14* **Typical taxiway markings and holding position sign**

NOTE Older holding points may be marked with white lines, but most are now yellow. Also, older holding points may still be marked with a single continuous line and a single broken line.

Markers on Unpaved Manoeuvring Areas

Aerodrome Boundary Markers

Orange/white striped wedge-shaped markers (like elongated wheel-chocks in shape), placed not more than 45 metres apart, indicate the boundary of an aerodrome. These are supplemented by flat orange/white markers, also placed 45 metres apart, on any structures which lie on the boundary.

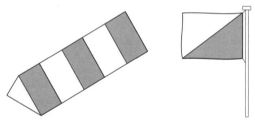

■ *Figure 3-15* **Boundary marker**

Unserviceable Portion

The orange/white striped wedge-shaped markers shown in Figure 3-15 are also used to mark the boundary of an unpaved area which is unserviceable for aircraft movement. These alternate with square flags showing equal orange and white triangular areas. Within this marked area the bad ground is itself marked with one or more white crosses (as described above).

Runway/Stopway Boundary Markers

White, flat rectangular markers, flush with the surface and placed not more than 90 metres apart, indicate the boundary of an unpaved runway or of a stopway. (A stopway is a prepared rectangular area of ground at the end of a runway, in the direction of take-off, designated as a suitable area in which an aircraft can be stopped in the case of an interrupted take-off.)

Light Aircraft Area

A white letter 'L' indicates a part of the manoeuvring area to be used only for the taking off and landing of light aircraft.

If a dumb-bell displayed in the aerodrome signals area has a red 'L' superimposed, it means that light aircraft are allowed to take off and land either on a runway or on the area designated by the white 'L'.

■ Figure 3-16 **Light aircraft areas**

Runway to be Used

A white 'T' (placed on the left side of a runway when viewed from the landing direction) indicates that *it* is the runway to be used. Where there is no runway it indicates the direction for take-off and landing.

■ Figure 3-17 **Runway to be used**

Landing Dangerous

A white cross displayed at each end of a runway indicates that landing is dangerous and that the aerodrome is used for storage purposes only.

■ Figure 3-18 **Landing dangerous**

Emergency Use Only

A white cross and a single white bar displayed at each end of the runway at a disused aerodrome indicates that the runway is fit for emergency use only. Runways so marked are not safe-guarded and may be temporarily obstructed.

■ *Figure 3-19* **Emergency use only**

Land in Emergency Only

Two vertical yellow bars on a red square on the signals area indicate that the landing areas are serviceable but the normal safety facilities are not available. Aircraft should land in emergency only.

■ *Figure 3-20* **Land in emergency only**

Displaced Threshold Markings

The threshold marking on a runway delineates the beginning of the usable portion of that runway (at the downwind end). Sometimes the threshold marking is moved, or displaced, some distance up the runway from the end of the paved area. Such a **displaced threshold** may be either temporary, to allow for maintenance, for instance, or permanent.

There are various displaced threshold markings, depending on the type (if any) of aircraft movement permitted in the first portion of the runway, and whether the displacement is temporary or permanent. Some pre-threshold areas may be usable for take-off, but not for landing; some may be unfit for any kind of aircraft movement. See overleaf.

Normal 'piano key' threshold marking for a paved runway.
'09' is the runway designator – the runway direction rounded off to the nearest 10°, 090° in this case

Permanently displaced threshold
White arrows indicate that the pre-threshold area is available for taxi and take-off, but not for landing

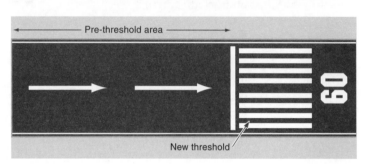

Permanently displaced threshold
White crosses indicate that the pre-threshold area is unfit for movement of aircraft and unsuitable as a stopway

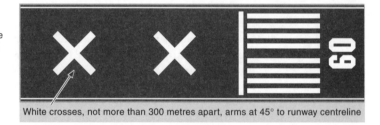

Temporarily displaced threshold
Pre-threshold area is available for taxi and take-off, but not for landing

Temporarily displaced threshold
Pre-threshold area is unfit for movement of aircraft and unsuitable as a stopway

■ *Figure 3-21* **Normal threshold marking (top) and displaced threshold markings**

Summary of Aerodrome Signals Visible only when on the Ground

In the Signals Area

1. A black ball on a mast signifies that the directions of take-off and landing are not necessarily the same.

2. Two red balls on a mast signify that gliding is taking place.

3. A rectangular red/yellow chequered flag or board means that aircraft may move on the manoeuvring area and apron only with the permission of ATC.

■ *Figure 3-22* **ATC in operation**

4. If the circuit direction at the aerodrome is variable, and a **left-hand circuit** is in operation, a **red flag** will be flown from the mast. A **green flag** on the mast signifies that a **right-hand circuit** is in force at the aerodrome. (Note that the colours of the flags for left and right circuits are the same as for aircraft navigation lights.)

■ *Figure 3-23* **Signals area at Wycombe (WP); clockwise from bottom left corner – special precautions; gliding in progress; (white dash symbol is part of dumb-bell not in use); right-hand circuits; and in centre: take-off and landing direction (towards the cross-arm)**

Away from the Signals Area

5. A square yellow board bearing a black 'C' indicates the position at which a pilot can report to ATC or other aerodrome authority.

■ *Figure 3-24* **Location of aerodrome authority**

Light Signals (Rule 46)

You should be aware of standard light signals that ATSU personnel may beam to aircraft. The light signals differ in meaning according to whether you are in flight or on the ground. Green flashes, for instance, when beamed at an aircraft in flight mean "Return for a landing", whereas when beamed to an aircraft on the ground they mean "Authorised to taxi".

FROM ATSU TO AIRCRAFT

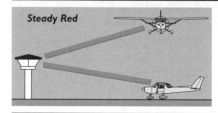

Steady Red

Do not land. Give way to other aircraft and continue circling.

Stop.

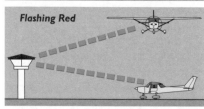

Flashing Red

Do not land. Aerodrome closed (go to another aerodrome).

Move clear of landing area.

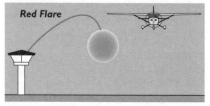

Red Flare

Do not land; wait for permission.

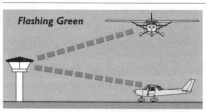

Flashing Green

Return to this aerodrome and wait for permission to land.

Cleared to taxi on the manoeuvring area if pilot satisfied no collision risk exists.

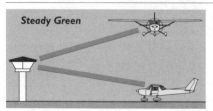

Steady Green

Cleared to land if pilot satisfied no collision risk exists.

Cleared to take-off if pilot satisfied no collision risk exists.

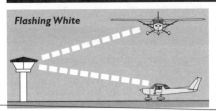

FROM ATSU TO AIRCRAFT

Flashing White

Land at this aerodrome after receiving a steady green light

Return to starting point on aerodrome.

Light signals can also be sent *from* an aircraft, but the equipment (such as flares) is rarely available for a pilot to use. The one signal that can be used in almost any aircraft, however, is flashing the landing lights or position navigation lights on and off (usually visible from the ground only at night) to indicate "I am compelled to land".

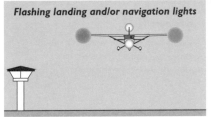

FROM AIRCRAFT TO ATSU

Flashing landing and/or navigation lights

I am compelled to land.

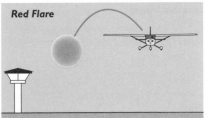

Red Flare

Immediate assistance required.

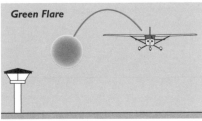

Green Flare

By night
May I land?

By day
May I land in a different direction from that indicated by the landing T?

Marshalling Signals (Rules 47 & 48)

FROM MARSHALLER TO PILOT (RULE 47)

Proceed under guidance of another marshaller

Right or left arm down, the other arm moved across body and extended to indicate position of the other marshaller.

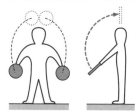

Move ahead

Arms repeatedly moved upward and backward, beckoning onward.

Open up starboard engine(s) or turn to port

Right arm down, left arm repeatedly moved upward and backward. The speed of arm movement indicates the rate of turn.

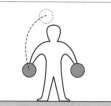

Open up port engine(s) or turn to starboard

Left arm down, the right arm repeatedly moved upward and backward. The speed of arm movement indicates the rate of turn.

Stop

Arms repeatedly crossed above the head. The speed of arm movement indicates the urgency of the stop.

FROM MARSHALLER TO PILOT (RULE 47)

Start engine

A circular motion of the right hand at head level, with the left arm pointing to the appropriate engine.

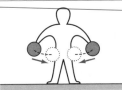

Chocks inserted

Arms extended, the palms facing inwards, then swung from the extended position inwards.

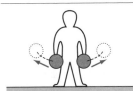

Chocks away

Arms down, the palms facing outwards, then swung outwards.

Cut engines

Either arm and hand placed level with the chest, then moved laterally with the palm facing downwards.

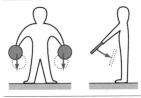

Slow down

Arms placed down, with the palms towards the ground, then moved up and down several times.

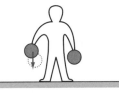

Slow down engine(s) on indicated side

Arms placed down, with the palms towards the ground, then either the right or left arm moved up and down indicating that the motors on the left or right side, as the case may be, should be slowed down.

FROM MARSHALLER TO PILOT (RULE 47)

This bay

Arms placed above the head in a vertical position.

Release brakes

Raise arm, with fist clenched, horizontally in front of the body, then extend fingers.

Engage brakes

Raise arm and hand, with fingers extended, horizontally in front of body, then clench fist.

All clear – marshalling finished

The right arm raised at the elbow with the palm facing forwards.

Start engine(s)

Left hand overhead with the number of fingers extended, to indicate the number of the engine to be started, and circular motion of right hand at head level.

Back aircraft tail to starboard

Point left arm down, move right arm down from overhead, vertical position to horizontal forward position, repeating right arm movement.

FROM MARSHALLER TO PILOT (RULE 47)

Back aircraft tail to port

Point right arm down, move left arm down from overhead, vertical position to horizontal forward position, repeating left arm movement.

 indicates that the signal is applicable only to helicopter operations.

 Hover

Arms placed horizontally sideways.

 **Land**

Arms placed down and crossed in front of the body.

 **Move upwards**

Arms placed horizontally sideways with the palms up beckoning upwards. The speed of the arm movement indicates the rate of ascent.

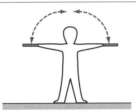 **Move downwards**

Arms placed horizontally sideways with the palms down beckoning downwards. The speed of arm movement indicates the rate of descent.

 **Move horizontally**

Appropriate arm placed horizontally sideways, then the other arm moved in front of the body to that side, in the direction of the movement, indicating that the helicopter should move horizontally to the left or right side, as the case may be, repeated several times.

FROM MARSHALLER TO PILOT (RULE 47)

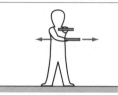

 Move back

Arms placed down, the palms facing forward, then repeatedly swept up and down to shoulder level.

 Release load

Left arm extended horizontally forward, then right arm making a horizontal slicing movement below left arm.

FROM PILOT TO MARSHALLER (RULE 48)

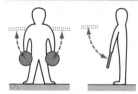

Brakes engaged

Raise arm and hand with fingers extended horizontally in front of face, then clench fist:

Brakes released

Raise arm with fist clenched horizontally in front of face, then extend fingers:

Insert chocks

Arms extended palms facing outwards, move hands inwards to cross in front of face:

Remove chocks

Hands crossed in front of face, palms facing outwards, move arms outwards:

FROM PILOT TO MARSHALLER (RULE 48)

Ready to start engines

Raise the number of fingers on one hand indicating the number of the engine to be started. For this purpose the aircraft engines shall be numbered in relation to the marshaller facing the aircraft, from his right to his left.

For example, No. 1 engine shall be the port outer engine, No. 2 shall be the port inner, No. 3 shall be the starboard inner, and No. 4 shall be the starboard outer.

Now complete **Exercises 3 – Aerodromes.**

Altimeter-Setting Procedures

Terminology

Atmospheric pressure decreases as altitude is gained. The altimeter, the flight instrument that is used to determine the vertical position of an aircraft, is a barometer that measures atmospheric pressure, but has a scale calibrated in feet rather than in units of pressure.

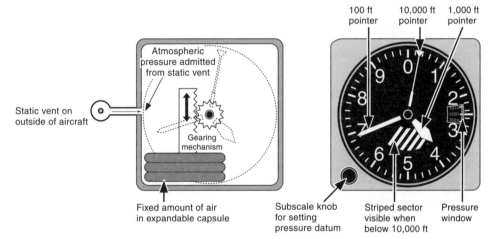

■ *Figure 4-1* **The altimeter is a pressure-sensitive instrument**

The pressure at the earth's surface changes from time to time and from place to place, so it is necessary to have some means of selecting the base level pressure from which height will be measured. To achieve this, altimeters have a **subscale** which can be altered with a knob – the 'subscale-setting knob'.

The pressure set on the altimeter subscale determines the pressure level from which the altimeter will indicate height. The usual pressure references are:

QNH. The *mean sea level* pressure at that time causing the altimeter to indicate **altitude** – the vertical distance above mean sea level (amsl). This is useful when separation from terrain is of concern, since the ground and obstacles are shown as heights amsl (i.e. elevations) on aeronautical charts.

QNH varies from time to time and from place to place as pressure systems move across the face of the earth, so the pilot needs to update the QNH on the altimeter subscale, both periodically and as he flies from region to region.

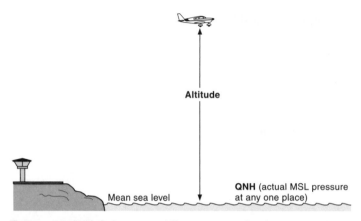

■ *Figure 4-2* ***Altitude is measured from mean sea level***

QFE. The pressure level at the aerodrome elevation (which is the highest point on the landing area) causing the altimeter to indicate **height** – the vertical distance above a specified datum, usually chosen to be above aerodrome level (aal). This is useful when the only concern is height above the aerodrome, and so QFE is a common setting to use for take-offs, circuits and landings.

QFE at an aerodrome will vary with time, information of any changes being passed to the pilot by ATC. QFEs at different aerodromes will differ depending upon elevation and the pressure pattern.

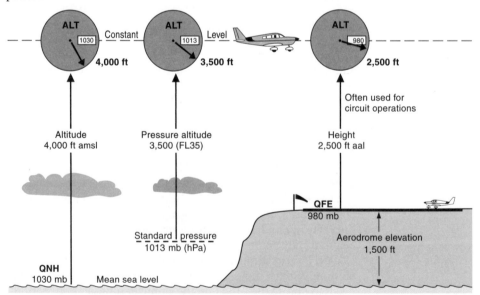

■ *Figure 4-3* ***Typical altimeter settings on a day of high pressure***

STANDARD PRESSURE. Setting the subscale to 1013.2 millibars (or hectopascals – see *Note* which follows) causes the altimeter to read **pressure altitude (PA)**. This is useful for vertical separation from other aircraft that are using the same altimeter setting. It is standard procedure to set 1013 at the higher cruising levels where terrain clearance is no longer a problem.

Using standard pressure avoids the need to update QNH with changes in time and/or location. To avoid confusing *pressure altitude* with *altitude*, pressure altitude generally has the last two zeros removed and is then referred to as a **flight level**, e.g. PA 3,500 is FL35.

NOTE You will recall that the term **hectopascal (hPa)** has been adopted as the international standard for the unit of atmospheric pressure; however, the UK has decided to continue using the unit **millibar**, at least for the time being. Hectopascals are used in most other countries. The unit size is identical, i.e. **1 mb = 1 hPa**. Below 5,000 ft, 1 mb equals approximately 30 feet.

Altimeter Setting Regions

The UK is divided into twenty **Altimeter Setting Regions** (ASRs) so that, when en route, all aircraft flying on QNH in the same region will have the appropriate Regional QNH set, allowing the pilots to ensure vertical separation between their aircraft. You must know which Altimeter Setting Region you are in, and when you are moving from one region to another. The Regional QNH is also known as the **regional pressure setting.**

The actual QNH will vary throughout the region, depending upon the pressure pattern. The Regional QNH is the lowest forecast QNH value for each hour, thereby ensuring that the pilot will be at, or slightly higher than, the altitude indicated.

Regional QNH is updated each hour by the Air Traffic Service (ATS) and as the pilot passes from one Altimeter Setting Region to another. No Aerodrome QNH in that region should be lower than the value of the Regional QNH. If desired, QNH values can be advised one hour in advance by ATC, any Meteorological Office or the Flight Information Service.

A chart showing all the UK Altimeter Setting regions appears in the UK Aeronautical Information Publication (AIP), ENR (En-Route) section, page 6-1-7-1. Figure 4-4 shows an excerpt from the chart.

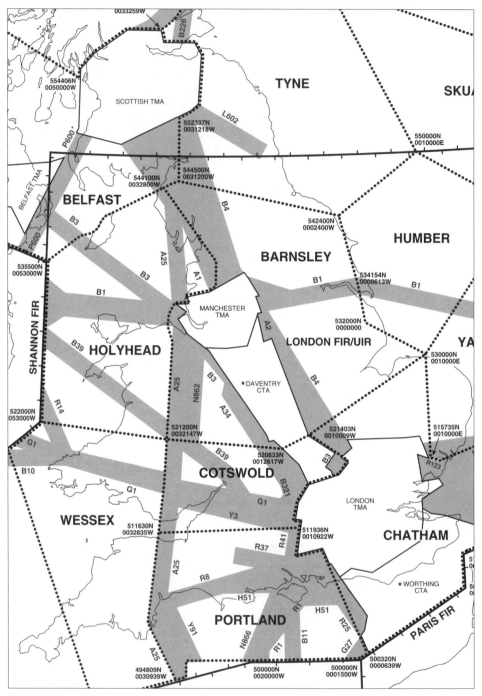

■ Figure 4-4 **UK Altimeter Setting Regions (ASRs)**

Vertical Separation from Terrain and Air Traffic

Altitude information is useful both for separation from terrain (since height on charts is given *above mean sea level*) and from other aircraft, so when flying reasonably low, QNH (sea level pressure) is set on the altimeter subscale as the reference datum.

The **transition altitude** is the altitude below which the vertical position of aircraft is controlled by reference to altitude – with pilots having QNH set on their altimeter subscales. Transition altitude is 3,000 ft amsl over most of the UK. Exceptions beneath certain airspace are listed in the UK Aeronautical Information Publication (AIP), ENR 1-7-1. This section also lists the UK transition altitudes.

When flying at a level well above the terrain, where separation between air traffic is the primary consideration, the standard pressure setting of 1013 mb is used as the reference datum.

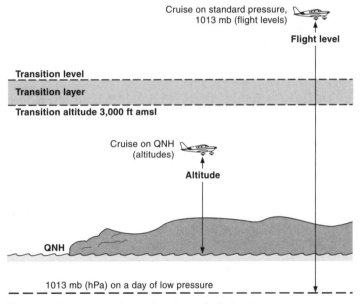

■ *Figure 4-5* **Transition from altitudes to flight levels**

Above the transition altitude, the term *flight level* is used, so standard pressure (1013) should be set on the altimeter subscale. This ensures safe vertical separation between Instrument Flight Rules (IFR) traffic. 1013 is used (subject to some exceptions, as specified in the UK AIP, ENR 1-7) by all aircraft in *controlled airspace*, or above 3,000 ft in *uncontrolled airspace*.

The **transition level** is the lowest flight level available for cruising above the transition altitude, the airspace between them being known as the *transition layer*.

Since the altimeter subscale setting will normally be changed when entering the transition layer, it is usual to express vertical displacement in terms of flight level after climbing through the transition altitude, and in terms of altitude after descending through the transition level.

In a flight plan:
- **altitudes** should be specified for that part of the flight below the transition altitude; and
- **flight levels** for that part of the flight above the transition level.

Cruising levels should be chosen:
- so that adequate terrain clearance is assured;
- so that any Air Traffic Service requirements are met; and
- to comply with the *quadrantal* or *semicircular* rules for cruise level, as specified for Instrument Flight Rules, if appropriate. See page 64.

A Typical Cross-Country Flight

Altimetry procedures are much easier in practice than it seems when you read the regulations. A typical cross-country flight between two aerodromes, remaining in uncontrolled airspace and probably not above 3,000 ft amsl, is illustrated below.

The main points are that:
- **Regional QNH is set en route** so that the altimeter indicates height above mean sea level.
- **In the circuit area** at each aerodrome, the altimeter subscale is set, according to the pilot's preference, to either:
  - **Aerodrome QFE,** so that the altimeter indicates height above aerodrome level (aal); or
  - **Aerodrome QNH,** so that the altimeter indicates height above mean sea level (amsl) or *altitude.*

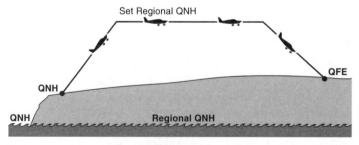

■ Figure 4-6 **A typical cross-country flight**

Take-Off and Climb

There are two altimeter readings that can be of value to a pilot for take-off and climb:

- ▣ **height above aerodrome level** (with QFE set) – useful for flying in the circuit, e.g. when a 1,000 ft circuit is flown 1,000 ft is indicated on the altimeter; and
- ▣ **altitude** (i.e. height above mean sea level, with QNH set) – useful for terrain and obstacle clearance.

WITHIN CONTROLLED AIRSPACE, where Air Traffic Control has a responsibility to separate air traffic, a pilot may use:

- ▣ either Aerodrome QNH or Aerodrome QFE for **take-off**. If the aircraft is fitted with two altimeters, one should be set to Aerodrome QNH to assist with vertical separation from other aircraft and terrain. (In other words, first altimeter set to Aerodrome QNH, and the second altimeter either to the same setting or on Aerodrome QFE.); and
- ▣ Aerodrome QNH during **climb** to, and while at or below, the transition altitude. Vertical position will be expressed as *altitude* on Aerodrome QNH.

After clearance to climb above the transition altitude has been given and the climb commenced, it is recommended that vertical position be expressed as a flight level, *provided* that the aircraft is not more than 2,000 ft below the transition altitude (unless specifically requested otherwise by ATC). Flight levels are based on *standard pressure* 1013 mb at all times and so are useful for vertical separation from other traffic using the same subscale setting. This occurs above the **transition level.**

OUTSIDE CONTROLLED AIRSPACE, a pilot may use any desired altimeter setting for take-off and climb (which, in practical terms, means either Aerodrome QNH or QFE). Vertical position should be reported as *altitude.* When under Instrument Flight Rules, however, vertical position must be expressed as a flight level (based on 1013) after climbing through the transition altitude.

Pilots taking off at aerodromes beneath Terminal Control Areas (TMAs) or Control Areas (CTAs) should use Aerodrome QNH when flying below the transition altitude and beneath these areas (to assist in vertical separation from them). However, Aerodrome QFE may be used within the circuit.

En Route

It is important to update the Regional QNH periodically, and when you enter a new altimeter setting region. If you are flying into an area of lower pressure, then the aeroplane will gradually descend if the original QNH is not altered.

Another effect results from a decreasing temperature because it increases the air density, the result being that a given pressure level is lower in cold air than in warm air. Thus, flying from a warm area to a cold area, the aeroplane will gradually descend.

In a very cold air mass, the aeroplane could be as much as 10% lower than the height indicated on the altimeter. The aeroplane should still be flown at a level according to the altimeter but, as the pilot-in-command, you should consider 'when flying from high to low, **beware below**'.

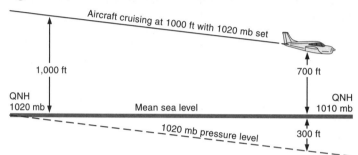

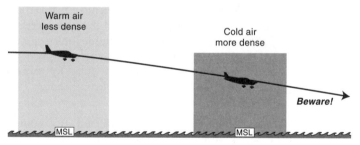

■ Figure 4-7 **When flying from high to low, beware below (valid for both low temperatures and low pressures)**

In areas of low pressure, the 1013 mb pressure level is *below* mean sea level and so FLs will be lower than at first thought. For example, if the QNH is 996 mb, the 1013 mb pressure level is almost 500 ft below mean sea level. FL40 will therefore be only 3,500 ft amsl.

If the intention is to be separated by 1,000 ft vertically from traffic or terrain at 3,000 ft amsl, then FL45 is the minimum suitable flight level, since it will (in this pressure situation) equate to an altitude of 4,000 ft.

A chart to convert altitudes to flight levels and vice versa is included at the end of the chapter.

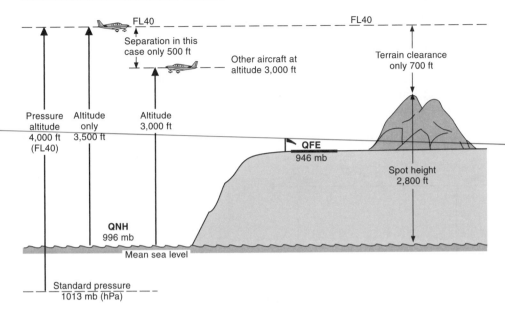

■ *Figure 4-8* **Select the minimum suitable flight level carefully, especially when the QNH is low**

Operations Within Controlled Airspace

Above 3,000 ft amsl, or above the appropriate transition altitude, whichever is the higher, en route aircraft are flown at *flight levels* (i.e. with standard pressure 1013 mb set), to assist in separation from other traffic.

Regional QNH is to be used en route, usually on the second altimeter, for checking terrain clearance. Aircraft flying in a Control Zone (CTR) or Terminal Control Area (TMA), at or below the transition altitude, will be given the appropriate QNH in their ATC clearance to enter the zone or area.

Operations Outside Controlled Airspace

AT OR BELOW 3,000 ft amsl, pilots may use any altimeter subscale setting. The most logical one to use when cruising is Regional QNH.

Pilots flying beneath a Terminal Control Area (TMA) or a Control Area (CTA), however, should use the QNH of an aerodrome situated beneath that area when flying below the *transition altitude.* This will not differ greatly from the Regional QNH and will be the same setting that aeroplanes above in the TMA or CTA are using. This will prevent inadvertent penetration of any controlled airspace above (due to any differences between local QNH and Regional QNH).

When penetrating a Military Air Traffic Zone (MATZ), the pilot will normally be given the Aerodrome QFE to enable vertical separation between aircraft. If there is more than one aerodrome in a combined MATZ (CMATZ), the lowest-value Aerodrome QFE will be given. This is known as the **clutch QFE**. However, in some circumstances when crossing a MATZ the QNH will be given.

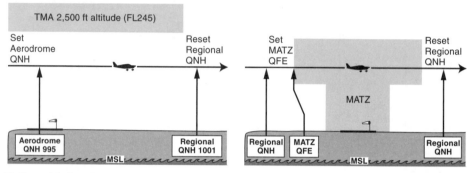

■ *Figure 4-9* **Aerodrome QNH (rather than Regional QNH) may be required en route**

AIC 9/2001 (Yellow 39) details procedures for penetration of a MATZ by civil aircraft. *Pooley's Flight Guide* also gives details on MATZ operations.

Outside controlled airspace, vertical position in flight plans and communications with the Air Traffic Service Unit is to be expressed as altitude. At or below 3,000 ft amsl on an Advisory Route, the Regional QNH should be set.

In level flight above 3,000 ft amsl (or above the appropriate transition altitude, whichever is the higher), pilots flying under the Instrument Flight Rules (IFR) must have 1013 mb set on an altimeter and conform to the **quadrantal rule** when selecting cruising levels. Vertical position will then be expressed as a *flight level*. Regional QNH should be used for checking terrain clearance.

VFR TRAFFIC ABOVE 3,000 ft amsl (or above the transition altitude), may operate on Regional QNH (for example, when the flight involves various headings and/or cruise heights); however, en route flights choosing to fly quadrantal levels *must* use the standard altimeter setting, 1013 mb.

The quadrantal rule and semicircular rule (Rule 30) is a means of vertically separating traffic flying in different directions. Whilst it is required for IFR traffic below FL245 (i.e. pressure altitude 24,500 ft), it is strongly suggested for VFR flights as well. The rule is based on magnetic track (rather than heading) since, in strong crosswind conditions, fast and slow aircraft maintaining identical tracks will have significantly different headings.

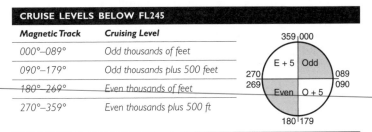

CRUISE LEVELS BELOW FL245		
Magnetic Track	*Cruising Level*	
000°–089°	*Odd thousands of feet*	
090°–179°	*Odd thousands plus 500 feet*	
180°–269°	*Even thousands of feet*	
270°–359°	*Even thousands plus 500 ft*	

■ *Figure 4-10* **The quadrantal rule for determining cruise levels below FL245**

EXAMPLE 1 For an aircraft *tracking* 330°M, a suitable level is FL45.

EXAMPLE 2 For an aircraft *heading* 085°M and experiencing 10° right drift, a suitable level is FL55 (since its magnetic track is 095°M).

ABOVE FL245, the semicircular rule applies, where vertical separation is increased because of the inaccuracies that are inherent in the altimeter.

CRUISE LEVELS ABOVE FL245		
Magnetic Track	*Cruising Level*	
000°–179°	*FL250, 270, 290, 330, 370*	
180°–359°	*FL260, 280, 310, 350, 390*	

■ *Figure 4-11* **The semicircular rule applies above FL245**

Approach and Landing

When an aircraft is descending from a *flight level* to an *altitude* preparatory to commencing approach for landing, ATC will pass the **Aerodrome QNH** (which may differ slightly from the Regional QNH, but should never be less than it).

On vacating the cruising flight level, the pilot will change to QNH unless further flight level reports have been requested, in which case the QNH will be set after the final flight level report is made. Thereafter the pilot will continue on the Aerodrome QNH until approaching the circuit area or established on final approach, when QFE or any other appropriate altimeter setting may be used. The only logical settings, of course, are either Aerodrome QNH or Aerodrome QFE.

On a **radar final approach** where the radar controller issues tracking and descent guidance, he will assume that an aircraft is using QFE and heights passed by the controller will be related to QFE (heights above the runway). A reminder of the assumed setting will be included in the radio phraseology.

To ensure safety it is recommended that all pilots use QFE, but if a pilot advises that he is using QNH, heights will be amended by the radar controller as necessary and *altitude* (height amsl) will be substituted for *height* in the radio calls.

Vertical positioning of aircraft during approach will, below transition level, be controlled by reference to altitudes (QNH) and then to heights (QFE). Pilots landing beneath Terminal Control Areas (TMAs) and Control Areas (CTAs) should use Aerodrome QNH when below the transition altitude and beneath these areas, except that the Aerodrome QFE may be used within the circuit area.

Missed Approach

On a missed approach (a go-around) pilots may continue with the altimeter setting selected for final approach, but reference to vertical position should be in terms of altitude on Aerodrome QNH, unless otherwise instructed by ATC.

Converting Between Altitudes and Flight Levels

A chart to convert altitudes to flight levels (and vice versa) is published in the UK Aeronautical Information Publication (AIP ENR 1-7-5). A reduced-size example of the conversion chart is reproduced in Figure 4-12.

To use this chart:
- take a vertical line upwards from the QNH reading along the bottom axis until it meets the appropriate (sloping) *flight level* line; then
- read horizontally across to the equivalent *altitude*.

Now complete **Exercises 4 – Altimeter-Setting Procedures.**

FLIGHT LEVEL GRAPH

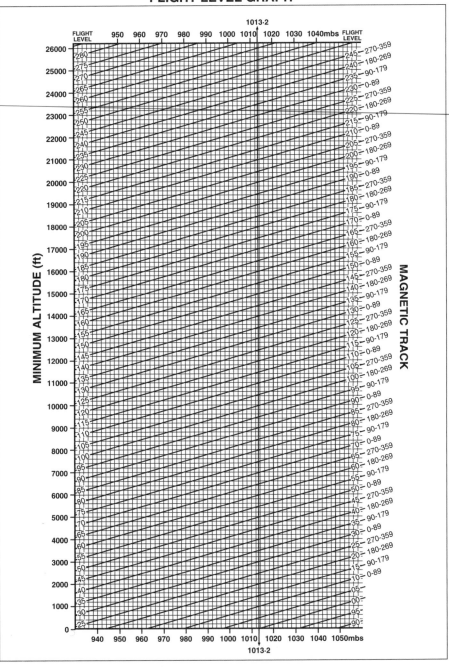

■ Figure 4-12 **The flight level–altitude conversion chart**

Airspace

Flight Information Regions

United Kingdom airspace has two Flight Information Regions – the London and Scottish Flight Information Regions (abbreviated FIR).

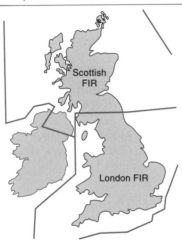

■ Figure 5-1 **London and Scottish Flight Information Regions (FIRs)**

The London and Scottish FIRs extend upwards from the surface to 24,500 feet (FL245). Above this the airspace is known as *upper airspace,* and abbreviated as UIR for *Upper Information Region,* i.e. the London and Scottish UIRs (from FL245 to FL660).

■ Figure 5-2 **Boundary between London and Scottish FIRs, as shown on the CAA 1:500,000 aeronautical chart**

Within the confines of the FIR structure, airspace is subdivided according to the amount and type of aeronautical activity which takes place in it. For example, the airspace around London is very busy, with a lot of commercial airline operations, business traffic, light aircraft on training and private flights, some military flights, plus gliders, balloons and microlights, etc. All this occurs in a quite small geographic area and so the airspace needs to be regulated closely to ensure safe use by the high volume of aircraft.

In contrast, the airspace over the Outer Hebrides needs little regulation due to the larger area and low density of traffic.

More comprehensive Air Traffic Services are provided in the busier airspace.

The Subdivision of Airspace

There are two distinct categories of airspace: **controlled airspace** and **uncontrolled airspace**. Within these two categories, various 'Classes' have been allocated to different parts, in line with a *classification system for civil airspace* recently introduced by the International Civil Aviation Organisation (ICAO). The new system is designed both to simplify airspace structure and to establish more commonality between countries.

The ICAO system grades airspace from A to G in order of importance. It begins with Class A, the highest status, which is allocated to the busiest controlled airspace. Classes B, D and E are allocated to other controlled airspace. Classes F and G cover all *uncontrolled airspace*. Specific VFR weather criteria and levels of Air Traffic Service apply to each airspace class. (Class C, although available in the ICAO system, is not currently allocated in the UK implementation of the ICAO system.)

Controlled Airspace

In controlled airspace, Air Traffic Control is provided to all flights. **UK controlled airspace** is made up of various aerodrome Control Zones (CTR), Terminal Control Areas (TMA), Control Areas (CTA) and Airways.

■ **A Control Zone (CTR)** is airspace around certain aerodromes in which Air Traffic Control (ATC) is provided to all flights. A Control Zone extends from ground level to a specified altitude or a specified flight level (FL), depending on height and has a minimum lateral dimension of 5 nm either side of the centre of the aerodrome in the direction of the approach path. The exceptionally busy London/Heathrow airport is situated in a Class A Control Zone (Class A is unavailable to VFR). However, most UK CTRs (such as Edinburgh, Newcastle, East Midlands, London/Gatwick, Belfast and Cardiff) are Class D Control Zones.

◻ **A Terminal Control Area** is a Control Area established at the confluence of controlled airspace routes in the vicinity of one or more major aerodromes. *Terminal Control Area* is sometimes abbreviated as *TCA,* but more commonly as **TMA** (from the earlier designation *Terminal Manoeuvring Area)*. There are currently four TMAs in the United Kingdom, with the busy London and Manchester TMAs allocated Class A, the Belfast TMA Class E, and the Scottish TMA Class D above 6,000 ft amsl and Class E at and below 6,000 ft amsl.

◻ **A Control Area (CTA)** is a portion of airspace in which Air Traffic Control is provided, and which extends upwards from a specified base altitude or flight level to an upper limit expressed as a flight level. The busy Control Areas in the UK (e.g. Cotswold, Daventry and Worthing CTAs) are Class A and the less busy ones (e.g. Birmingham, Luton and Stansted CTAs) are Class D airspace.

◻ **An Airway** is a Control Area in the form of a corridor and is delineated by radio navigation aids. Each Airway has an identification code (e.g. A25 or *Alpha Two Five,* R8 or *Romeo Eight),* and extends 5 nm each side of a straight line joining certain places, with specified vertical limits.

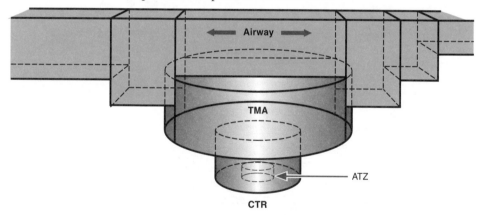

■ *Figure 5-3* **A Control Zone (CTR), Terminal Control Area (TMA) and an Airway**

All Airways are Class A except where they pass through a TMA, CTA or CTR of lower status. They are used by airliners (and other Instrument Flight Rules traffic) travelling between the principal aerodromes. As it approaches an aerodrome, the lower level of an Airway is usually stepped down to provide controlled airspace protection for air traffic on climb and descent.

As Class A airspace is unavailable to VFR flights, Class B is only upper airspace (above FL245), and Class C is not currently allocated in the UK, VFR operations in controlled airspace in the UK will usually be confined to Classes D and E.

Depiction of Controlled Airspace on Charts

Controlled airspace up to FL245 (24,500 ft), and its classification, is depicted on the ICAO 1:500,000 (half-million) aeronautical chart series published by the CAA. The CAA 1:250,000 (quarter-million) series shows all controlled airspace with a base at or below 3,000 ft amsl.

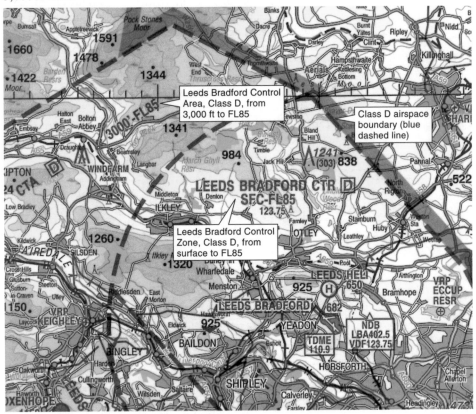

■ Figure 5-4 **Part of the Leeds Bradford CTA and CTR shown on a CAA 1:250,000 chart)**

A fuller description of the boundaries and vertical extent of controlled airspace (CTRs, CTAs, TMAs and Airways) is given in the UK Aeronautical Information Publication (AIP) in the ENR section. Also, *Pooley's Flight Guide* contains a useful list of Controlling Authorities and Communications Channels for all UK controlled airspace.

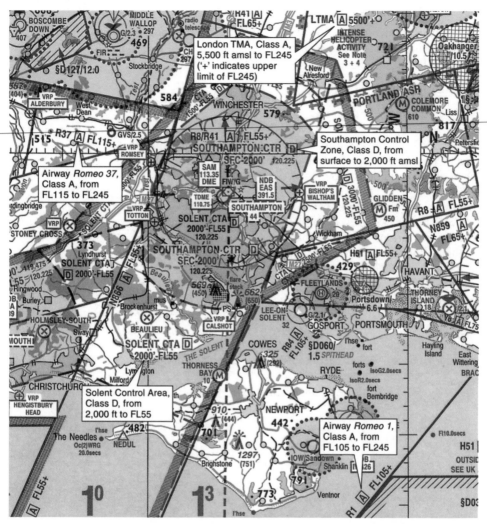

■ *Figure 5-5* **Controlled airspace and its classification is shown on CAA 1:500,000 aeronautical charts**

Classification of Airspace

Each class of airspace available to VFR operations has specific Visual Meteorological Conditions (VMC) criteria, which are expressed in terms of minimum flight visibility and distance from cloud. However, ANO Schedule 8, which concerns the privileges of the Private Pilot's Licence (and others) may be more stringent than the VFR minima and must be observed by the non-holder of an IMC or Instrument Rating. This is covered in detail in Chapter 7, *Visual Flight Rules (Rules 24–27)*, and Chapter 10, *Pilots' Licences*.

The Special VFR Clearance (SVFR)

Special VFR is an authorisation to fly within a Control Zone:

- although the pilot is unable to comply with Instrument Flight Rules (IFR);
- at night;
- to transit Class A airspace of a lower status such as the London and Channel Islands CTRs;
- when weather is below VMC minima.

Special VFR is designed to facilitate flights that would otherwise be restricted due to weather and the clearance may be given when traffic conditions allow (as determined by the appropriate ATC unit).

Some UK aerodromes have Special VFR Routes and/or Entry/Exit Access Lanes established with published minimum weather conditions (flight visibility and, in some cases, cloud ceiling) applicable to the route or lane. These are listed in AIP AD 2-22 and *Pooley's Flight Guide*.

Note that the issue of a 'Special VFR clearance' by ATC to fly in a Control Zone does not absolve the pilot from the responsibility to observe the Rules of the Air, where applicable. For example, if a Special VFR clearance was accepted to fly across London at 1,000 ft in the London CTR it might cause an aircraft to contravene the 'land clear' requirements of Rule 5 of the Rules of the Air.

Entry/Exit Lanes

a   To permit aircraft to operate to and from Edinburgh Airport in IMC but not under IFR the following entry/exit lanes have been established for use, under the conditions stated, as follows:

    i   1   a lane 3 nm wide, known as the Polmont Lane, with centre-line the M9 Motorway extending from Grangemouth (near the western boundary of the Edinburgh Control Zone) eastwards, via the Polmont Roundabout, Linlithgow Loch and Philpstoun to a point at which it joins the Edinburgh Aerodrome Traffic Zone;

        2   a lane 3 nm wide, known as the Kelty Lane, with centre-line the M90 Motorway extending from Kelty (near the northern boundary of the Edinburgh Control Zone) southwards, across the Forth Road Bridge, to a point at which it joins the Edinburgh Aerodrome Traffic Zone;

    ii   use of the lanes is subject to clearance being obtained from ATC Edinburgh, irrespective of prevailing weather conditions. This clearance is to be obtained by non-radio equipped aircraft before take-off and by radio equipped aircraft before entering the lane;

    iii   aircraft using the lanes must remain clear of cloud and in sight of the ground or water, not above 2000 ft (Edinburgh QNH), and in flight visibility of not less than 3 km;

    iv   an aircraft using a lane shall keep the centre-line on its left, unless otherwise instructed by ATC for separation purposes. In these circumstances ATC will pass traffic information to the aircraft concerned;

    v   pilots of aircraft are responsible for maintaining adequate clearance from the ground or other obstacles.

b   Additionally, to permit the effective integration of traffic, flights operating in VMC and under VFR may be required by ATC to follow these routes as detailed in paragraph 8.

■ *Figure 5-6* **AIP AD listing for access to Edinburgh under SVFR: clear of cloud, in sight of surface, not above 2,000 ft and minimum flight visibility of 3 km**

Further information on Special VFR is contained in AIP AD 2, ENR 1-2 and *Pooley's Flight Guide*.

Crossing an Airway

Crossing the Base of an Airway

A basic PPL holder (i.e. no IMC or Instrument Rating) may fly at right-angles across the base of an en route section of an Airway where the lower limit is defined as a flight level (and not an 'altitude', i.e. height amsl), but must not enter the Airway. This is only in weather conditions of at least VMC. For example, if the lower limit of an Airway running east–west is FL75, then the pilot may cross it by flying north or south not above FL75.

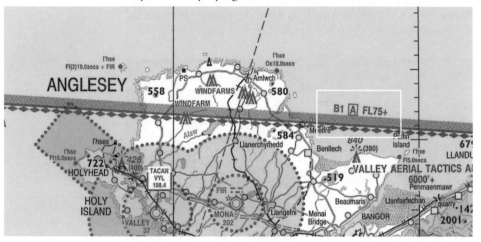

■ Figure 5-7 **Airway B1 (as shown on the half-million chart) can be crossed at right-angles at FL75 in VMC**

Penetrating an Airway

A pilot who holds a valid Instrument Rating, even in an aeroplane not fully equipped for IFR flight, may penetrate and cross an Airway in Visual Meteorological Conditions (VMC) by day provided he:

- ▢ files a flight plan either before departure or when airborne; and
- ▢ requests a *crossing clearance* from the responsible Air Traffic Control unit when at least 10 minutes from the intended crossing point, and subsequently receives that ATC crossing clearance.

Except where otherwise authorised by ATC, aircraft are required to cross an Airway by the shortest route (normally at right-angles) and to be in level flight at the cleared level on entering the Airway.

Aerodrome Traffic Zones (ATZ) in Control Zones

Aerodrome Traffic Zones exist at many aerodromes within UK Control Zones. ATZs are also established at most aerodromes outside controlled airspace, and we cover them in detail on page 78.

Uncontrolled Airspace

Uncontrolled airspace in the UK includes Advisory Routes and Open-FIR. Open-FIR includes various areas and zones (covered shortly). As far as Air Traffic Control Services are concerned, the advisory service provided in uncontrolled airspace is less comprehensive than the control service provided in controlled airspace.

Advisory Routes (ADR)
Advisory Routes are allocated Class F. Although deemed to be 10 nm wide (the same as Airways), Advisory Routes are depicted as centrelines on the CAA half-million aeronautical charts by blue dashed lines.

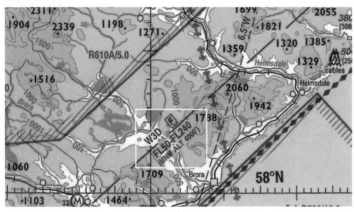

■ Figure 5-8 **Advisory route W3D (Whiskey three delta) shown on the Scotland 1:500,000 chart**

VFR operations on an Advisory Route require a flight plan to be lodged. En route, these flights then receive an Air Traffic Advisory Service from the Air Traffic Service Unit (ATSU) responsible for the route (refer AIP ENR 3 and *Pooley's Flight Guide*).

Open-FIR
Open-FIR, *Class G* airspace, covers about 50% of all UK airspace. This is all previously unallocated airspace, and includes Radar Advisory Service Areas (RASAs), Military Aerodrome Traffic Zones (MATZs), and the Aerodrome Traffic Zones (ATZs) which are located at most UK aerodromes outside controlled airspace.

Air Traffic Services in Open-FIR
■ A **Flight Information Service (FIS)** is available to all aircraft in both UK FIRs through "London Information" and "Scottish Information" Air Traffic Services Units. Air traffic services provided to flights in Open-FIR include:
  – information and warnings on meteorological conditions;

- changes of serviceability in navigational and approach aids;
- condition of aerodrome facilities;
- alerting service for known aircraft in need of search and rescue;
- aircraft proximity warnings;
- other advisory information pertinent to the safety of air navigation.

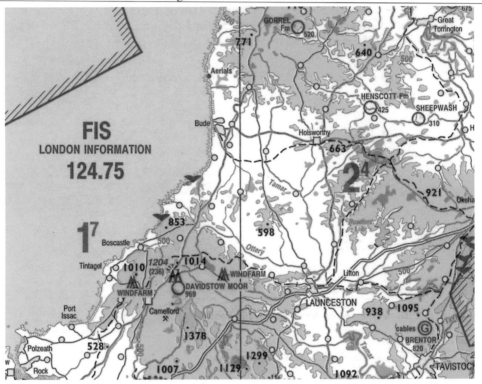

■ *Figure 5-9* **Open-FIR on a 1:500,000 chart; information (London) frequency is 124.75 MHz in this area**

■ **Radar services** are also available in Open-FIR: either a *Radar Advisory Service (RAS)* or a *Radar Information Service (RIS)* provided by various Air Traffic Service Units, including those participating in the **Lower Airspace Radar Advisory Service (LARS)** – explained further in the next chapter.

■ In **ATZs** and **MATZs** discrete communication frequencies are established for the responsible aerodrome authorities.

Certain Open-FIR is Regulated Airspace

Flight operations in Aerodrome Traffic Zones are regulated by Rule 39 of the Rules of the Air. Consequently, along with controlled airspace, ATZs are *regulated airspace*.

Aerodrome Traffic Zones (ATZs)

Aerodrome Traffic Zones operate at most UK civil and military aerodromes. ATZs are not allocated a specific airspace classification, but adopt the class of the airspace within which they are located.

The standard physical dimensions of an ATZ are:

- from ground level to 2,000 ft above aerodrome level (aal);
- within the area bounded by a circle of radius:
  - (i) 2 nm, where length of longest runway is 1,850 m or less; or
  - (ii) 2.5 nm, where longest runway is greater than 1,850 m;
  - – the centre of the circle being the mid-point of the longest runway.

NOTE Some 2 nm ATZs are expanded to 2.5 nm radius to provide at least 1.5 nm clearance from the end of all runways.

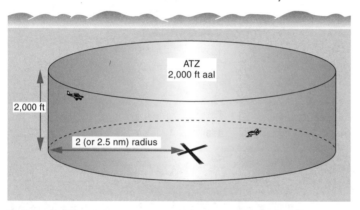

■ *Figure 5-10* ***Dimensions of an Aerodrome Traffic Zone (ATZ)***

Rules to Enable Safe Flight in an ATZ (Rule 39)

An aircraft must not fly within an Aerodrome Traffic Zone unless the pilot-in-command:

- has the permission of the appropriate Air Traffic Control (ATC) unit; or
- where there is no ATC unit, he has obtained sufficient information from the Aerodrome Flight Information Service (AFIS) unit to enable flight within the zone to be made safely; or
- where there is no ATC or AFIS unit, he has obtained information from the Air/Ground (A/G) radio station at the aerodrome to enable the flight to be made safely.

Aircraft flying in an ATZ must:

- maintain a continuous watch on the appropriate radio frequency notified for communications at the aerodrome or, if this is not possible, keep a watch for visual instructions;

■ where the aircraft has radio, give position and height to the aerodrome ATC or AFIS unit or A/G radio station, as the case may be, on entering and leaving the zone; and

■ make any other standard calls or requested calls.

These requirements apply at:

■ a Government aerodrome at such times as are notified (usually H24, i.e. continuous);

■ an aerodrome having an ATC or AFIS unit, during the hours of watch;

■ a licensed aerodrome having A/G radio communication with aircraft, during the hours of watch, and whose hours of availability are detailed in AIP AD 2.

NOTE When an ATZ is established in a class of airspace having more stringent *rules* than those of an ATZ (e.g. Class D airspace), then the more stringent rules take precedence.

Depiction of Aerodrome Traffic Zones

ATZ locations and details (radius, hours of watch and radio frequency) are listed in AIP AD 2, and any amendments notified by NOTAM (Notice to Airmen). They are also shown on new UK 1:500,000 and CAA 1:250,000 aeronautical charts; however, ATZs located wholly in regulated airspace are not shown on 1:500,000 charts to avoid congestion on the chart. *Pooley's Flight Guide* lists aerodrome operating hours; in the main these coincide with ATZ hours.

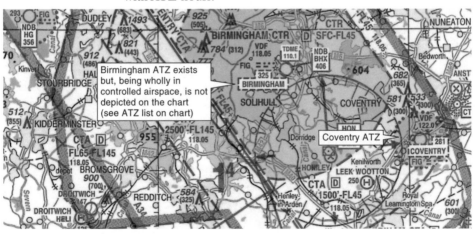

■ *Figure 5-11* **ATZs outside regulated airspace are shown on the 1:500,000 chart**

CAA 1:500,000 charts have a full list of Aerodrome Traffic Zones printed on the left-hand side, along with the Air/Ground frequency of the responsible radio unit at the aerodrome.

AERODROME TRAFFIC ZONE (ATZ), is regulated airspace from the surface to 2000ft AAL within a circle centred on the notified mid-point of the longest runway, radius 2·0NM (RW≤ 1850m) or 2·5NM (RW≥1850m), where Mandatory Rules apply.
Most Government Aerodrome ATZs are H24.

ABERPORTH (EGUC)	AFIS 122·15
ANDREWSFIELD (EGSL)	AFIS or A/G 130·55
BARKSTON HEATH (EGYE)	INITIAL CALL ATC CRANWELL 119·375
BECCLES (EGSM)	A/G 134·6
BEMBRIDGE (EGHJ)	AFIS or A/G 123·25
BENSON (EGUB)	ATC 120·9
* BIGGIN HILL (EGKB)	ATC 129·4
♦ BIRMINGHAM (EGBB)	ATC 118·05
BLACKBUSHE (EGLK)	INITIAL CALL FARNBOROUGH/ODIHAM APP 125·25, AFIS or A/G 122·3
BODMIN (EGLA)	A/G 122·7
BOSCOMBE DOWN (EGDM)	ATC 126·7
BOURN (EGSN)	A/G 129·8
♦ BOURNEMOUTH (EGHH)	ATC 119·625
♦ BRISTOL (EGGD)	ATC 128·55
♦ BRIZE NORTON (EGVN)	ATC 119·0
CAERNARFON (EGCK)	ATC VALLEY 134·35, A/G 122·25
* CAMBRIDGE (EGSC)	ATC 123·6 or 122·2
♦ CARDIFF (EGFF)	ATC 125·85
CHICHESTER /Goodwood (EGHR)	AFIS 122·45
CHIVENOR (EGDC)	A/G 130·2
CLACTON (EGSQ)	A/G 135·4
COLERNE (EGUO)	ATC 122·1
COLTISHALL (EGYC)	ATC 125·9
COMPTON ABBAS (EGHA)	A/G 122·7
CONINGSBY (EGXC)	ATC 120·8
COSFORD (EGWC)	ATC 128·825
COTTESMORE (EGXJ)	ATC 130·2
* COVENTRY (EGBE)	ATC 119·25
* CRANFIELD (EGTC)	ATC 122·85
CRANWELL (EGYD)	ATC 119·375
CULDROSE (EGDR)	ATC 134·05

■ *Figure 5-12* **Example list of ATZs from a CAA 1:500,000 chart**

Military Air Traffic Zones (MATZ)

A Military Aerodrome Traffic Zone (MATZ) is specified airspace surrounding many UK military aerodromes:

☐ from the surface up to 3,000 feet above aerodrome level within a radius of 5 nm; and usually

☐ with a stub (or stubs), width 4 nm, extending out a further 5 nm along final approach path(s) for the main instrument runway(s) between 1,000 and 3,000 ft above aerodrome level (aal).

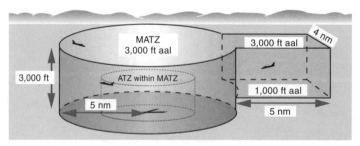

■ *Figure 5-13* **Dimensions of a typical Military Air Traffic Zone (MATZ)**

Pilots of civil aircraft wishing to penetrate a Military Air Traffic Zone are strongly advised to do so under the control of the MATZ ATC authority, in accordance with published procedures. These are detailed in AIP ENR 2-2 and *Pooley's Flight Guide*.

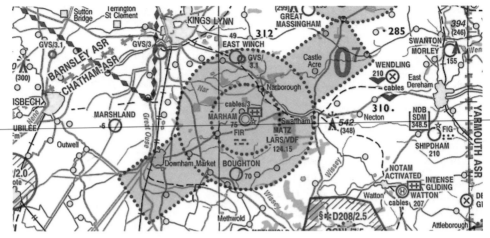

■ *Figure 5-14* **Marham MATZ, as shown on the 1:500,000 aeronautical chart**

NOTE Even if you should choose to ignore the MATZ airspace (which technically is non-regulated), there is, within the MATZ, an Aerodrome Traffic Zone (ATZ) which must be observed at all times (most are 'H24', i.e. continuous).

Summary of UK Airspace Classes

Controlled Airspace

CLASS A is allocated to the busiest airspace – Airways (except where they pass through a TMA, CTA or CTR of lower status), London (Heathrow) CTR, Channel Islands CTR/CTA, London and Manchester TMAs, and Cotswold, Daventry and Worthing CTAs. **Class A is not available to VFR flights.**

CLASS B airspace – allocated to upper airspace, i.e. above FL245 (24,500 ft), and so not usually of concern to VFR pilots.

CLASS C airspace – not currently allocated in UK (although details of Class C airspace are included in the AIP ENR 1 *Airspace Classifications* table, because parts of the Republic of Ireland are shown on some charts and all Irish controlled airspace below FL200 is Class C).

CLASS D airspace – allocated to less-busy controlled airspace (most UK CTRs and CTAs) and the Scottish TMA above 6,000 ft amsl, including any Aerodrome Traffic Zones (ATZ) in the Class D Control Zones.

VFR flights in Classes B and D airspace require appropriate flight notification to be given and an ATC clearance obtained.

CLASS E airspace – only allocated to the Scottish CTR (including ATZs in the Control Zone), the Scottish TMA at and below 6,000 ft amsl and the Belfast TMA.

Class E airspace is similar to Class D but differs from Class D in not requiring flight notification or ATC clearance, and in having a reduced traffic information service from ATC.

Uncontrolled Airspace

CLASS F airspace – Advisory Routes. Use of Class F airspace requires lodging a flight plan, and aircraft receive an Air Traffic Advisory Service from the responsible ATS unit.

CLASS G airspace – all previously unallocated airspace, and called 'Open-FIR'. Class G includes Radar Advisory Surveillance Areas (RASAs), Military Aerodrome Traffic Zones (MATZs), the Aerodrome Traffic Zones (ATZs) located at aerodromes outside controlled airspace (see Rule 39).

Flight Information Service and Radar Services in Class G airspace are available.

VMC Criteria

Minimum weather conditions for VFR operations in the various Classes of airspace are considered in Chapter 7, *Visual Flight Rules*.

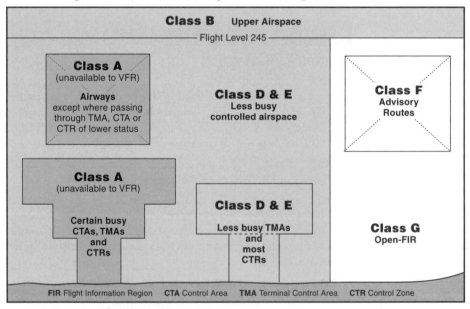

■ *Figure 5-15* ***Summary of UK airspace classifications***

Flight Restrictions and Hazards

The UK Aeronautical Information Publication (AIP) contains the **Chart of United Kingdom Airspace Restrictions and Hazardous Areas** and a tabulation of the full details for each area depicted. Amendments are notified by NOTAM (Notice to Airmen) and AIC (Aeronautical Information Circular).

Aeronautical Charts also depict Airspace Restrictions and Hazards, the manner of presentation varying slightly with the chart used. Each chart will have a **legend** listing the various features and how they are depicted. Some chart *legend notes* also refer to hazardous areas.

Specific types of restriction or hazard of interest to the private pilot are given below.

Prohibited, Restricted and Danger Areas

Prohibited areas

A Prohibited Area is defined airspace in which flight is prohibited.

■ *Figure 5-16* **Prohibited Area P611, surface up to 2,200 ft (1:500,000 chart)**

Restricted Areas

A Restricted Area is defined airspace in which flight is restricted according to certain conditions.

Danger Areas

A Danger Area is defined airspace in which activities dangerous to flight may occur. They are specified in the ENR 5 section of the UK Aeronautical Information Publication (AIP) and on the **Chart of Airspace Restrictions and Hazardous Areas**, where they are shown as:

- ■ **Solid red** outline if they are active in published hours (i.e. scheduled);
- ■ **Pecked red** outline if they are inactive unless notified by NOTAM.

A Danger Area designated as D129/1.75 indicates that it is:

- located between 51°N and 52°N latitudes, as shown by the first number (a '0' would indicate a Danger Area that is just North of 50°N, and a '6' that it is just North of 56°N); and
- extends from the surface to 1,750 ft amsl (as shown by the '1.75'; '1–2.5' would indicate a Danger Area that extends from 1,000–2,500 ft amsl).

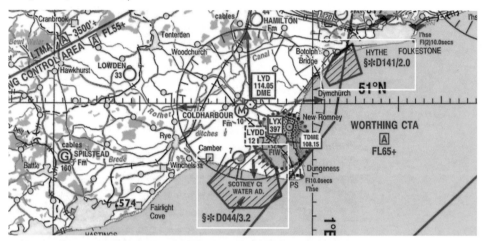

■ *Figure 5-17* **Danger Areas D044 and D141 (1:500,000 chart)**

- **On other aeronautical charts** Danger Areas may be depicted differently, and you should refer to the *legend* on the particular chart in use.

Danger Area Services available for certain Danger Areas are shown on CAA half-million (1:500,000) charts. These are the:
- **Danger Area Crossing Service (DACS)**; and
- **Danger Area Activity Information Service (DAAIS)**.

The chart listings include contact frequencies for the units providing Air Traffic Service in the particular Danger Area. *Pooley's Flight Guide* also contains listings of the units providing the DACS and DAAIS.

NOTE Danger areas with identification numbers prefixed by an asterisk (★) contain airspace subject to by-laws which prohibit entry during the period of activity. See UK AIP ENR 1-1.

Some Danger Areas in the UK are known as *Weapon Range Danger Areas* (WRDA), and details are covered by a *legend note* on CAA 1:500,000 (half-million) charts. It is strongly recommended that aircraft flying in or near these areas obtain a radar service.

> DANGER AREA CROSSING SERVICE (DACS) is available for certain Danger Areas. The relevant areas (identified on the chart by the prefix †) and Unit Contact Frequencies to be used are shown below. For availability of the services see UK AIP ENR 5.1.
>
> | | |
> |---|---|
> | D001 | ST MAWGAN APP 126·5MHz* |
> | D003,D004,D006A,D007,D007A,D007B,.... | { PLYMOUTH MIL 121·25MHz OUTSIDE TIMES |
> | D008, D008A, D008B, D009 & D009A | { LONDON MIL VIA LONDON INFO 124·75MHz* |
> | D006 | CULDROSE APP 134·05MHz* |
> | D012, D013, D014, D017, | { PLYMOUTH MIL 124·15MHz OUTSIDE TIMES |
> | D021, D023, & D031 | { LONDON MIL VIA LONDON INFO 124·75MHz |
> | D036 | PLYMOUTH MIL 124·15MHz SEE UNDER LIST |
>
> DANGER AREA ACTIVITY INFORMATION SERVICE (DAAIS) is available for certain Danger Areas shown on this chart (identified by the prefix §). The Nominated Air Traffic Service Units (NATSUs) to be used are shown below. See UK AIP ENR 1.1. Pilots are advised to assume that a Danger Area is active if no reply is received from the appropriate NATSU.
>
> | | |
> |---|---|
> | D037, D038, D039, D040 | LONDON INFORMATION 124·75MHz/ 124·6MHz |
> | ✳D044 | LYDD INFORMATION 120·7MHz** |
> | D060 | SOLENT APP 120.225MHz |
> | D061 | EXETER APP 128·15MHz |
> | ✳D131, D132, D133 & D133A | FARNBOROUGH APP 125·25MHz** |
> | D136, D138, D138A & D138B | SOUTHEND APP 128·95MHz** |
> | ✳D141 | LYDD INFORMATION 120·7MHz** |

■ *Figure 5-18* **Excerpt of DACS and DAAIS listings on the Southern England 1:500,000 aeronautical chart**

For further information on Prohibited, Restricted and Danger areas, see AIP ENR 5.

Pilotless Aircraft

Pilotless aircraft operate and manoeuvre in certain Danger Areas. The aircraft are orange or red and may be flown by day and night in all weather conditions. These aircraft may operate up to 60,000 ft, under radar control.

Details of the Danger Areas concerned are given in AIP ENR 5-1 (primarily West Wales). Pilots should note that, because pilotless aircraft are less manoeuvrable in the landing configuration than manned aircraft, care should be taken to avoid that part of Danger Areas in which recovery of pilotless aircraft takes place. This can be ascertained by contacting the controlling authority.

Air Navigation Obstructions

Details of structures which reach a height of **300 feet above ground level** (agl) are published in AIP ENR 5-4. They are also shown on certain aeronautical charts. Obstructions which are over **500 ft agl** are lit. Those between 300 and 500 ft agl may or may not be lit.

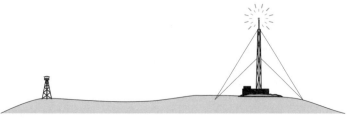

■ *Figure 5-19* **Air navigation obstructions may, or may not, be lit**

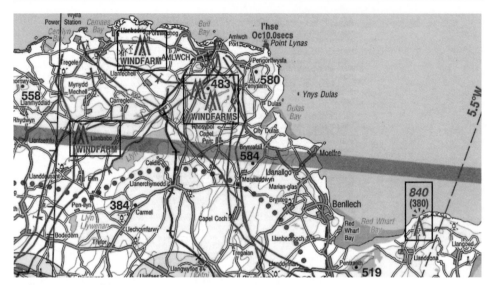

■ *Figure 5-20* **Unlit obstructions (windfarms) and a lit obstruction of height 840 ft amsl, 380 ft agl (1:250,000 chart)**

Aerodrome Obstructions

Obstructions on or near aerodromes and which are considered hazardous to aircraft landing or taking off are listed in the AD section of the UK AIP and are shown on Instrument Approach and Landing charts.

Glider Launching Sites

Glider launching may take place from aerodromes or other sites. Pilots are warned that a winch-launched glider may carry the cable up to a height of 2,000 ft agl before releasing it. At a few sites this height may be exceeded. In addition, *aerotows* are common at gliding sites, in which case the glider and tug aircraft will be linked by a cable, the combined length of which may extend to a maximum of 150 m horizontally.

Parachute Training Balloons

Captive balloons used for military parachute training may be flown without notification. They will not be flown above 1,000 ft agl, and will be flown in a horizontal visibility of not less than 5 km and a separation from cloud of not less than 1,000 ft.

Parachuting Sites

Regular free-fall parachuting up to FL150 takes place at the sites and during the periods listed in the UK AIP. Some Government and licensed aerodromes where regular parachuting occurs are included in the list, but parachuting may also take place during daylight hours at any Government or licensed aerodrome.

All parachutists are required to operate only in weather conditions which will enable them to remain clear of cloud, in sight of the surface and in a flight visibility of at least 5 km. A list of parachute dropping zones (DZs) and relevant telephone numbers and ATC frequencies for DZ activity is detailed in AIP ENR 5-5.

■ *Figure 5-21* **Free-fall parachuting site on 1:250,000 chart. Parachutists may be expected within a circle of 1.5 nm radius of the drop zone (DZ), up to FL150.**

Parascending

Parascending (where a parachutist is towed aloft by a winch and cable) also takes place at certain locations throughout the UK.

Because of the low-speed characteristics of parascenders, and the difficulty in seeing them in certain conditions, they are a potential hazard to other airspace users. Known parascending sites are therefore listed in the UK AIP.

Parascenders in thermals may be at heights up to the cloud base or base of controlled airspace, whichever is lower. Launch cables may be carried up to 2,000 ft agl.

Cable Launching Activities

Sites where launching by winch and cable of gliders, hang-gliders and parascending parachutes is permitted are listed in AIP ENR 5 and shown on the 1:500,000 topographical chart. Note that the cable may be carried to 2,000 ft agl, and in some cases, even higher.

Hang-Gliding Sites

Thermalling hang-gliders may be encountered at heights up to the cloud base or the base of controlled airspace, whichever is the lower. Known hang-gliding sites are listed in AIP ENR 5-5 and *Pooley's Flight Guide.*

At certain sites hang-gliders may be launched by winch or auto-tow, and cables may be carried up to 2,000 ft agl.

Microlight Flying Sites

Areas at which intensive flying by microlight aeroplanes may take place are listed in AIP ENR 5-5 and *Pooley's Flight Guide,* and are also shown on aeronautical charts (see Figure 5-22).

■ *Figure 5-22* **Microlight flying sites (Swinford and Saddington) and glider launching sites (Rothwell and Lyveden) on a 1:500,000 chart**

Areas of Intense Aerial Activity (AIAA) and Military Low Flying

The UK AIP gives details of areas of intense air activity and the military low flying system. Pilots should be extremely vigilant when operating in Areas of Intense Aerial Activity, keeping a very good lookout and making use of any radar service available. Avoid AIAAs if possible, entering them only if no other suitable route is available.

Military low flying occurs in many parts of the UK at heights up to 2,000 ft agl. As the **greatest concentration** is between 250 and 500 ft agl, civil pilots are strongly recommended to avoid this height band whenever possible.

Gas Venting Sites

The UK AIP ENR 1.1 lists gas venting sites together with advisory minimum altitudes for overflight. Severe turbulence may be experienced near these sites when gas venting is in progess and light aircraft pilots are strongly advised to avoid flying over such areas where possible. Gas venting sites (GVS) are indicated on charts by a magenta circle and the hazard altitude is shown in thousands of feet.

Small Arms Ranges

Ranges with a vertical hazard height of 500 ft agl do not attract UK Danger Area status, however, firing at some ranges may constitute a hazard to aircraft operating below 500 ft agl. For details see AIP ENR 5.3 and Chart of UK Airspace Restrictions (ENR 6-5-1-1).

Target Towing Trials

Aircraft towing targets on up to two thousand feet of cable may be flown in daylight, in Visual Meteorological Conditions (VMC) only, up to a height of 10,000 ft. The areas in which these trials take place are given in the UK AIP.

Bird Hazards and Sanctuaries

Bird concentration areas are shown in AIP ENR 6, together with the location of bird sanctuaries. Pilots should avoid flying below the listed *effective altitude* over bird sanctuaries and below 1,500 ft agl over any other areas where birds are known (or likely) to concentrate. Where it is essential to fly lower, bear in mind that the risk of a bird strike increases with speed.

AIP ENR 6 gives a general description of seasonal migration patterns, along with charts depicting *Bird Concentration Areas*.

Permanent warnings are published (AIP ENR 6 and *Pooley's Flight Guide*) for certain aerodromes where birds habitually present a hazard at or near the surface. **Temporary warnings** are issued by ATS Units for aerodromes and as 'Temp Nav Warnings' (TNW) when there are indications that migration or mass flocking of birds is imminent.

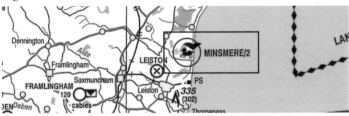

■ Figure 5-23 **Minsmere bird sanctuary on the 1:500,000 chart.**
Bird sanctuaries are also shown on 1:250,000 charts.

High Intensity Radio Transmissions

AIP ENR 5 lists certain sources of high-intensity radio and radar transmissions. Flight within the specified High Intensity Radio Transmission Areas (HIRTA) may lead to interference with communication and navigational equipment, and flight within some areas could endanger health.

Royal Flights

A 'Royal Flight' is a civil or military flight over the UK carrying one or more of the principal members of the Royal Family. The special conditions applicable to such flights vary depending on whether the Royal Flight is by fixed-wing or helicopter:

- **Fixed-wing**: Royal Flights are conducted where possible within existing controlled airspace. Where this is not possible temporary controlled airspace (known as 'Temporary Class A Airspace') is established along the route. Irrespective of the weather conditions, all controlled airspace used (whether existing or temporary) is classified as *Class A* airspace (i.e. *Instrument Flight Rules only*). Details of Temporary Class A Airspace are promulgated by NOTAM and a Freephone telephone service is available on 0500-354802. Information on Royal Flights and Temporary Restricted Airspace (such as for the *Red Arrows*) is updated daily. Ensure that you are fully briefed on these important matters before becoming airborne. For further information see AIC 101/2002 (Yellow 95) and AIP ENR 1-1 and the AIS website at www.ais.org.uk (see page 11 of this manual).

- **Helicopter**: No special ATC procedures are employed, but details are promulgated by NOTAM. You should keep a sharp lookout for the aircraft and keep well clear. In the event of getting into close proximity, however, the normal right-of-way rules apply.

*Now complete **Exercises 5 – Airspace**.*

Air Traffic Services

ICAO Annex 11 – Air Traffic Services

Annex 11 contains the international standards and recommended practices relating to the establishment of airspace, air traffic services units and air traffic services necessary to expedite the safe, orderly flow of air traffic.

Objectives and Need for ATS

2.2 The objectives of the air traffic services shall be to:

- prevent collisions between aircraft in the air and on the ground;
- prevent collisions between aircraft and objects on the manoeuvring area of aerodromes.
- expedite and maintain an orderly flow of traffic;
- provide advice and information useful for the safe and efficient conduct of flights;
- notify and cooperate with appropriate organisations regarding aircraft in need of search and rescue.

2.4 The need for the provision of air traffic services shall be determined by considering at least the following:

- types of air traffic involved;
- density of air traffic;
- meteorological conditions.

Classification of Airspaces

We have already seen how airspace is classified. The Annex contains details of the air traffic services relating to each class of airspace as follows:

2.6 ATS airspaces shall be classified and named as follows:

- **Class A.** IFR flights only permitted; all flights are subject to ATC service and are separated from each other.
- **Class B.** IFR and VFR flights; all flights are subject to ATC service and are separated from each other.
- **Class C.** IFR and VFR flights; all flights subject to ATC service. IFR flights are separated from other IFR flights and from VFR flights. VFR flights are separated from IFR flights and receive traffic information about other VFR flights.
- **Class D.** IFR and VFR flights; all flights subject to ATC service. IFR flights are separated from other IFR flights and receive traffic information in respect of VFR flights. VFR flights receive traffic information about all other flights.

☐ **Class E.** IFR and VFR flights; all flights subject to ATC service. IFR flights are separated from other IFR flights. All flights receive traffic information as far as is practical.

☐ **Class F.** IFR and VFR flights; all participating IFR flights receive an air traffic advisory service and all flights receive flight information service if requested.

☐ **Class G.** IFR and VFR flights; all flights receive flight information service if requested.

Air Traffic Services in the UK

There are many Air Traffic Service Units (ATSUs) spread throughout the UK, whose function it is to assist the passage of aircraft by providing information prior to flight and in maintaining radio contact during flight.

Area Control Centre (ACC) Services

An Area Control Centre (ACC) provides the following air traffic services:

☐ an **Air Traffic Control Service** to aircraft operating on Airways (IFR flights);

☐ an **Air Traffic Advisory Service** to aircraft flying on Advisory Routes;

☐ a **Flight Information Service** (FIS) and Alerting Service;

☐ a **Distress and Diversion service.**

Aerodrome Traffic Services

Air Traffic Control (ATC)

ATC at an aerodrome is responsible for the **control** of aircraft in the air in the vicinity of the aerodrome, and all traffic on the manoeuvring area. All movements on the manoeuvring area are subject to the prior permission of ATC.

The total Air Traffic Control responsibility is shared between **Aerodrome Control** and **Approach Control**. Aerodrome Control is responsible for aircraft on the manoeuvring area except runways in use.

The point dividing the responsibilities of Aerodrome Control and Approach Control for aircraft on runways in use and in the air may vary with weather conditions or other considerations, but it is the normal rule that departing aircraft contact Aerodrome Control first and arriving aircraft contact Approach Control first.

Non-radio aircraft may be given instructions or information by **lamp or pyrotechnics** from the control tower and by **ground signals** in the aerodrome signals area. Lamp and pyrotechnic signals may be made to any aircraft, radio-equipped or otherwise, from a subsidiary control point such as a runway control van.

Flight Information Service (FIS)

FIS is provided at some aerodromes which do not have ATC to give information relevant to the safe and efficient conduct of flights in the ATZ. The Flight Information Service is not permitted to give instructions at any time as FIS is not a *control* service but an *information only* service. When FIS is provided the callsign suffix "Information" is used, e.g. "Rochester Information".

Air/Ground Radio Stations

There are some aerodromes which provide neither Air Traffic Control (ATC) nor Flight Information Service (FIS) but which do have an Air/Ground (A/G) radio station through which aircraft and the aerodrome authority can communicate. It should be kept in mind that such facilities are usually operated by people without ATC qualifications. The callsign suffix "Radio" is used to distinguish them, for example "Nottingham Radio".

Air Traffic Services in Open-FIR

The Lower Airspace Radar Service (LARS)

When flying in UK unregulated airspace up to and including FL95 within the limits of radio/radar coverage, pilots are recommended to use the Lower Airspace Radar Service (LARS). LARS is a structured national system whereby nominated civil and military ATSUs, if their primary duties permit, may provide the Radar Advisory Service (RAS) or the Radar Information Service (RIS), on request, within approximately 30 nm of each unit. If three attempts to establish radio contact with a LARS station are made without success, it may be assumed that the station is not operating.

THE RADAR ADVISORY SERVICE (RAS) is available on request from a pilot. The radar controller will use radio communications to provide:

- **traffic information** (bearing, distance and level if known); and
- **advisory avoiding action** necessary to maintain separation from the other aircraft.

A service will only be provided to flights under Instrument Flight Rules (IFR) irrespective of meteorological conditions. (Yes, in most circumstances, you may fly under IFR in the UK provided you remain in VMC conditions.)

Controllers will expect the pilot to accept radar vectors or level changes that may require flight into IMC conditions. Therefore, if any doubt exists that the flight can be continued in VMC, pilots *not* qualified to fly in IMC should not accept a RAS.

VFR pilots must not accept an instruction that would take them into cloud. The radar controller should be told immediately of any instructions (heading or level changes) that are not acceptable to the pilot, in which case the pilot becomes responsible for his own avoiding action. Any self-initiated changes in heading or altitude by the pilot should be immediately advised to the radar controller, since it may affect separation from other aircraft.

THE RADAR INFORMATION SERVICE (RIS) is available on request from a pilot. The radar controller will use radio communications to provide:

■ traffic information only (bearing, distance and level if known – no avoiding action will be offered).

This service is useful in good VMC conditions, where the radar controller can provide the pilot with an extra pair of eyes, without repeated avoiding action which may be unnecessary in the good conditions and time-wasting for the flight. The pilot should advise the radar controller if he is changing level, level-band or route. A RIS may be offered if RAS is impracticable.

Both RAS and RIS are mainly utilised through LARS units but are also available from some other ATSUs, on request by the pilot. Remember that, even if in receipt of a Radar Advisory Service outside controlled airspace, a PPL holder is still responsible for maintaining terrain clearance, maintaining at least the required visibility minima according to his licence, avoiding collisions, and obtaining permission to enter Class D controlled airspace and an Aerodrome Traffic Zone.

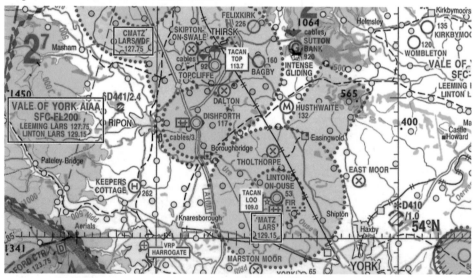

■ Figure 6-1 **LARS frequencies are shown on the 1:500,000 and 1:250,000 charts**

THE FLIGHT INFORMATION SERVICE (FIS) existed before the advent of the radar services and is still available. The service provides local information on weather, serviceability of radio navaids, aerodrome conditions and other reported traffic.

THE PROCEDURAL SERVICE is a non-radar ATC service mainly used to separate IFR traffic by time and/or distance using an approach control service, and for aircraft flying along advisory routes.

Pre-Flight Briefing Services

Aeronautical Information Services (AIS) are available to pilots preparing for a flight. Most UK aerodromes are equipped with self-briefing documents where pilots can obtain latest information which may affect their flights. British Isles En Route and Aerodrome Bulletins, together with Navigation Warning Bulletins and NOTAM information, are provided for VFR flights at the aerodromes where self-briefing facilities exist.

The CAA has introduced a facsimile and screen-based service to provide information to Private Pilots on Aerodromes, Navigational Warnings, Royal Flights and En Route Aids, plus details of Changes and Restrictions to Airspace. For more information, see AIP GEN 3-1.

The Flight Plan

The *flight plan* is an ATC message, compiled by, or on behalf of, the aircraft commander (pilot-in-command) to a set ICAO format and then transmitted by the appropriate ATS authority to organisations concerned with the flight. It is the basis on which an ATC clearance is given for the flight to proceed.

Correct use of the CAA Flight Plan Form (form CA48) is essential – particularly in these days of automatic data processing. Incorrect completion may well result in a delay in processing and consequently a delay to the flight. Full instructions for completion of the Flight Plan Form are contained in the CAA General Aviation Safety Sense Leaflet No. 20 *Instructions for the Completion of Flight Plan Forms* and AIP ENR 1-10. See also Vol. 3 of *The Air Pilot's Manual* series.

Note that a pilot intending to make a flight must in any case contact ATC (or other authority where there is no ATC) at the aerodrome of departure. This is known as **booking out** and is a separate and additional requirement to that of filing a flight plan.

Also note that where an aerodrome is notified as **Prior Permission Required** (PPR) the filing of a flight plan does *not* constitute prior permission (see page 31).

Private pilots may file a flight plan for any flight. They are advised to file a flight plan if intending to fly more than 10 nm from the coast or over sparsely populated or mountainous areas.

All pilots *must* file a flight plan for flights:

- **within Class A airspace;**
- **within controlled airspace** in Instrument Meteorological Conditions (IMC) or at night, excluding Special VFR (SVFR);
- **within controlled airspace** in Visual Meteorological Conditions (VMC) if the flight is to be conducted under the Instrument Flight Rules (IFR);
- **within Class D** Control Zones and Control Areas, irrespective of flight rules –VFR or IFR (This requirement may be satisfied by passing flight details on R/T.);
- **within the Scottish and London** Upper Information Regions (UIRs), i.e. above FL245 (Class B airspace);
- **where the destination** is more than 40 km (21.6 nm) from departure and maximum total weight authorised exceeds 5,700 kg;
- **to or from the UK** which will cross a UK Flight Information Region (FIR) boundary;
- **during which it is intended** to use the Air Traffic Advisory Service on an Advisory Route (ADR).

NOTE The fact that **night flying** is conducted in accordance with IFR procedures does not, of itself, require that a flight plan be filed. Equally, IFR flight in the Open-FIR, by day or night, does not, of itself, require a flight plan.

Normally flight plans should be filed at least 30 minutes before requesting taxi or start-up clearance (60 minutes in certain cases where the controlling authority is London, Manchester or Scottish Control). If this is not possible, a flight plan can be filed when airborne. If intending to enter controlled airspace, at least 10 minutes' notice must be given.

If your departure airfield does not have an Air Traffic Services Unit (ATSU), flight plans may be submitted by telephone to designated air traffic units (normally at major airports). A responsible person should be nominated to inform the 'parent' ATSU once the flight is airborne of the time of departure. Similarly, a responsible person needs to be nominated at an arrival airfield without an ATSU. This is to ensure that alerting action will be taken if the aircraft fails to arrive.

If your 'off-blocks' (chocks) flight-planned departure time is delayed by more than 60 minutes, the flight plan should be cancelled and a new flight plan submitted (30 minutes if part of the flight enters controlled airspace).

If a pilot who has filed a flight plan lands at an aerodrome other than the destination specified, the Air Traffic Services Unit at the specified destination must be advised within **30 minutes** of the estimated time of arrival.

Meteorology

The primary method of obtaining a pre-flight meteorological briefing in the United Kingdom is by **self-briefing**, using either:

- facilities, information and documentation routinely available or displayed in aerodrome briefing areas;
- the AIRMET automated telephone service, the dial-up METFAX facsimile-based service for pilots and the UK Meteorological Office's PC-based MIST (Meteorological Information Self-briefing Terminal) service; or
- the Met Office's website at www.met-office.gov.uk.

Details of the various methods of obtaining weather information are set out in a handy pocket-sized booklet called *Get Met*, which is published jointly by the CAA and the Met Office and is available free of charge from the Met Office (fax: 0845 3001300; e-mail: metfax@meto.gov.uk).

Self-briefing does not require prior notification by the pilot. Where necessary, the personal advice of a forecaster or additional weather information can be obtained from the designated Forecast Office for the departure aerodrome (see AIP GEN 3-5).

AIRMET weather information is provided by the Met Office and made available to aerodrome briefing offices and to individual users. Methods of dissemination to individual users utilise the following:

- the public telephone service (voice messages, spoken at dictation speed); and
- the METFAX Service (text information).

AIRMET information comprises:

- low-level Area Forecasts issued four times daily for three regions covering the UK and near continent (Scottish, Northern and Southern England); and
- Aerodrome Forecasts (TAFs) and Actual Weather Reports (METARs) for over 100 UK and near-continent aerodromes.

The AIRMET TAFs and METARs are available via the AIRMET automated telephone service (accessed using a tone-dialling telephone equipped with a key pad, or suitable equivalent), and via the dial-up METFAX Service. (They are also available from Forecast Offices and at aerodrome briefing offices.) A menu of codes enables the caller to select the required information by keying in the code after the service answers. For more information, refer to AIP GEN 3-5.

METFAX (the facsimile-based MET service) can be accessed using any modern facsimile equipment. A large range of products is available; this is detailed on the information sheet which is available on the service.

The weather information covers flights within the UK and near Continent and consists of:

■ **Area Forecasts** of weather between the surface and 15,000 ft amsl, with winds and temperatures up to at least 18,000 ft;

■ **Terminal Aerodrome Forecasts** (TAFs) and/or **Aerodrome Reports** (METARs).

In addition to the Met Office, there are a number of commercial websites that provide weather information, but none of them should be be relied on in your pre-flight plannning.

NOTE Should you experience severe weather conditions (particularly if they were not forecast) you should advise the appropriate ATC unit by radio as soon as possible.

For flights outside the area of coverage of the above area forecast systems, *special forecasts* are provided by the designated Forecast Offices, as listed in the UK AIP (GEN 3-5). Requests for such forecasts should include details of the route, the period of the flight, the height to be flown and the time at which the forecast is to be collected.

At least **4 hours'** notice should be given for flights over 500 nm, and at least **2 hours'** for shorter flights. At less notice the best possible service will be given in the time available, but it may be only a briefing, without documentation. A UK area forecast relevant to the flight will also have to be obtained.

Further details of areas, telephone numbers and procedures are given in AIP GEN 3-5 and *Pooley's Flight Guide.*

Regular **VOLMET** broadcasts of weather reports and tendency at certain aerodromes are made on specific VHF frequencies.

Broadcasts of recorded information for certain aerodromes are made on selected VHF-NAV radio frequencies or on discrete VHF-COM radio frequencies to relieve congestion on Air Traffic Control communication frequencies. This is known as the **Automatic Terminal Information Service** (ATIS). At larger airports the ATIS may be split into arrivals and departures, each broadcast on separate frequencies.

ATIS frequencies, where the service is available, are listed in the AD section of the AIP and in the aerodrome directory of the *Pooley's Flight Guide.*

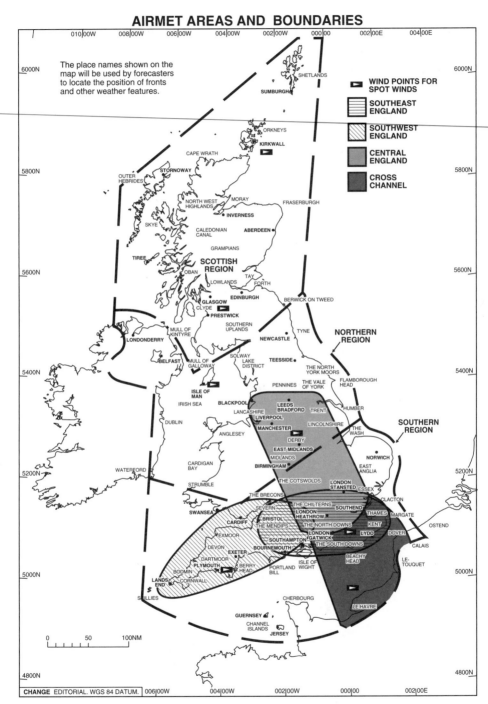

■ Figure 6-2 **AIRMET boundaries and regions (see Chapter 26)**

Facilitation

Facilitation is the simplification of formalities in moving aircraft across international boundaries.

Private Aircraft on International Flights

Since the completion of the European Single Market on 1 January 1993, Customs and Excise requirements vary according to whether or not the place of departure or the destination is within the European Union (EU).

Arrival or Departure for UK outside the EU

Broadly speaking, unless special permission has been obtained from Customs and Excise, the commander of an aircraft arriving or departing the UK, from or to places outside the EU, may not do so at any place other than a designated Customs and Excise airport as listed in the UK AIP, GEN 1. For full particulars see the AIP GEN 1-2.

Arrival or Departure for UK inside the EU

Since 1993 formalities for the departure and arrival for UK have been significantly relaxed and non-designated aerodromes (even private airstrips) may be used.

However, due to the revision of the Prevention of Terrorism Act 2000, any person who has arrived or is seeking to leave Great Britain by aircraft from a non-designated airfield must give a minimum of 12 hours notice in writing (24 hours in some regions) to the relevant police department. Facsimile is usually acceptable.

Breaches of these provisions may result in prosecution, which is punishable upon conviction by up to 3 months imprisonment, a fine of £2,500 or both.

Procedures may vary between aerodromes and advice should be sought from individual aerodrome operators as well as AIP GEN 1-2.

Flights within the Common Travel Area (Channel Islands, Republic of Ireland, Isle of Man, Northern Ireland), will normally require 24 hours' notice of arrival or departure (because of the requirements of the Prevention of Terrorism Act 2000), if your departure or arrival is from a non-designated airfield. Effectively the permission of Special Branch has to be obtained. Again, facsimile is normally acceptable.

NOTE The Channel Islands are regarded as places outside the EU for the purposes of arrival and departure procedures. See AIP GEN 1-2.

Eurocontrol En Route Navigation Services

Charges for en route navigation services (i.e. other than those provided in connection with the use of an aerodrome) in the London and Scottish FIRs and UIRs may be levied in respect of private aircraft. Certain flights are exempted, notably those:

- made entirely under VFR;
- starting and finishing at the same aerodrome;
- made by aircraft with a maximum total authorised weight of less than 2,000 kg.

Now complete **Exercises 6 – Air Traffic Services.**

Visual Flight Rules (Rules 24–27)

Visual reference to the outside environment both for attitude reference and to navigate the aeroplane is necessary for the visual pilot (i.e. the pilot holding a basic PPL, without either an IMC Rating or an Instrument Rating).

If visual reference is lost – for example, by inadvertently entering cloud or by flying in conditions of reduced visibility – the results may be disastrous. To avoid this, minimum flight visibility requirements and minimum distances from cloud have been established. These are known as the **Visual Flight Rules (VFR)**.

The starting point for the Visual Flight Rules is:
- **a minimum flight visibility** of 8 km; and
- **a minimum distance from cloud** of 1,500 metres horizontally and 1,000 ft vertically.

Under certain conditions these requirements are reduced, e.g. below FL100 (10,000 ft) the flight visibility minimum is reduced to 5 km.

Conditions in which flight is possible under the Visual Flight Rules are known as **Visual Meteorological Conditions (VMC)**.

VMC Minima for Airspace Classes
Each airspace class (except Class A, which is unavailable to VFR flights) has VMC minima specified for VFR flights. These VMC minima are specified in Rules 25 and 26 of the Rules of the Air.

VMC in Controlled Airspace (Rule 25)

VMC IN CONTROLLED AIRSPACE	
Class A	*VFR flights not permitted*
Class B	*Only allocated to upper airspace in the UK, i.e. at and above FL245, so not of significance to most VFR flights (VMC minima in Class B airspace are 8 km flight visibility and clear of cloud).*
Class C	*Not allocated in the UK, but same as Class D below (for those planning to fly in Irish controlled airspace).*
Class D	■ *Above FL100, minimum flight visibility: 8 km.* ■ *Below FL100, minimum flight visibility: 5 km.* ■ *Minimum distance from cloud: 1,500 metres horizontally and 1,000 ft vertically.*
Class E	*same as Class D*

An exemption to Rule 25 permits VFR flight in UK Class D and E airspace at and below 3,000 ft amsl at an IAS of 140 knots or less when clear of cloud, in sight of the surface and in a flight visibility of at least 5 km. Helicopters may fly at and below 3,000 ft amsl clear of cloud and in sight of the surface. This facilitates such flight operations as circuit training and other low-level, low-speed VFR operations at aerodromes located within Class D and E airspace.

NOTE Should a flight be 'cleared' to maintain VMC and its own separation it is the pilot's responsibility to do so.

VMC outside Controlled Airspace (Rule 26)

VMC OUTSIDE CONTROLLED AIRSPACE	
Class G	■ *Above FL100, minimum flight visibility 8 km*
	■ *Below FL100, minimum flight visibility 5 km*
	■ *Minimum distance from cloud: 1,500 metres horizontally and 1,000 ft vertically*
	But:
	When flying at or below 3,000 ft amsl, the aeroplane must fly clear of cloud, with a minimum flight visibility of 5 km. This can be reduced to 1,500 metres minimum flight visibility when flying at speeds of 140 kt IAS or less (restricted to 3 km for non-IMC or non-instrument-rated holders).
Class F	*same as Class G*

Outside controlled airspace, helicopters may operate in conditions less than 1,500 m flight visibility, provided that their forward speed allows sufficient collision-avoidance time in the prevailing conditions.

NOTE Under the Visual Flight Rules, the pilot is solely responsible for the safety of the flight, separation from other aircraft, terrain clearance, and for remaining at a satisfactory distance from cloud in adequate flight visibility.

A table of airspace classifications, published on the *Chart of Airspace Classifications* (AIP ENR 6-1-4-1), includes a summary of the UK VMC minima. This is reproduced in Figure 7-1.

VFR Operations

In interpreting the Visual Flight Rules, it is essential to remember that a pilot must at all times fly within the privileges of the licence which is held.

In the UK, VFR flights must always be operated in conditions of flight visibility and separation from cloud which at least meet the minimum legal requirements of two considerations:

1. **The VMC minima** for the class of airspace; and
2. **The privileges of your licence** (as laid down in ANO Schedule 8). It is important that you do not confuse VFR flight minima with the privileges of the Private Pilot's Licence, which *may* be more restrictive.

UK ATS AIRSPACE CLASSIFICATIONS

CONTROLLED AIRSPACE

	A	C	D	E	F	G
IFR — SEPARATION	All aircraft	IFR from IFR, IFR from VFR	IFR from IFR	IFR from IFR	IFR from IFR (participating IFR traffic)	Not provided (See Note)
IFR — SERVICES	Air traffic control service	Air traffic control service	Air traffic control service including traffic information about VFR flights (and traffic avoidance advice on request)	Air traffic control service and traffic information about VFR flights as far as practical	Air traffic advisory service, Flight information service	Flight information service (See Note)
IFR — SPEED LIMITATION	Not applicable +				250kt IAS below FL100	250kt IAS below FL100
IFR — RADIO	[radio]	[radio]	[radio]	[radio]	Not required	Not required
IFR — ATC CLEARANCE	Required	Required	Required	Required	Not required	Not required
VFR — SEPARATION	VFR FLIGHT NOT PERMITTED	VFR from IFR	Not applicable	Not provided	Not provided (See Note)	Not provided (See Note)
VFR — SERVICES		1) Air traffic control service for separation from IFR 2) VFR traffic information (and traffic avoidance advice on request)	Air traffic control service - traffic information on all other flights	Traffic information as far as practical	Flight information service (See Note)	Flight information service (See Note)
VFR — VMC MINIMA		# 8km — 1500m (FL100, 1000ft); 5km — 1500m (below FL100); 3000ft AMSL 140kt 5km or less OR clear of cloud, in sight	# 8km — 1500m (FL100, 1000ft); 5km — 1500m (below FL100); 3000ft AMSL 140kt 5km or less OR clear of cloud ‡, in sight	# 8km — 1500m (FL100, 1000ft); 5km — 1500m (below FL100); 3000ft AMSL 140kt 5km or less OR clear of cloud ‡, in sight	# 8km — 1500m (FL100, 1000ft); 5km — 1500m (below FL100); 3000ft AMSL 5km ✳ OR clear of cloud, in sight	# 8km — 1500m (FL100, 1000ft); 5km — 1500m (below FL100); 3000ft AMSL 5km ✳ OR clear of cloud, in sight
VFR — SPEED LIMITATION		250kt IAS below FL100	250kt IAS below FL100	250kt IAS below FL100	250kt IAS below FL100	250kt IAS below FL100
VFR — RADIO		[radio]	[radio]	Not required	Not required	Not required
VFR — ATC CLEARANCE		Required	Required	Not required	Not required	Not required

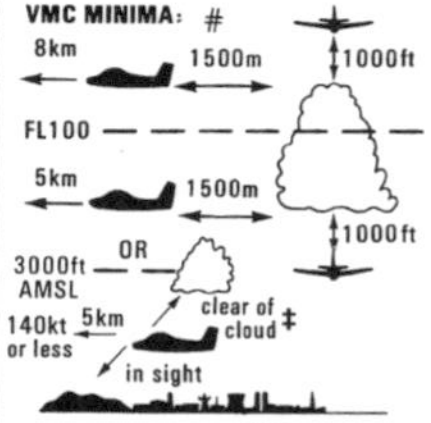

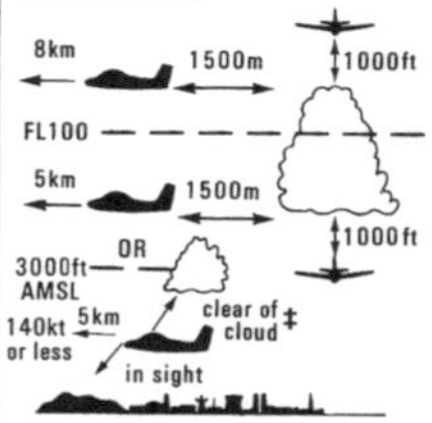

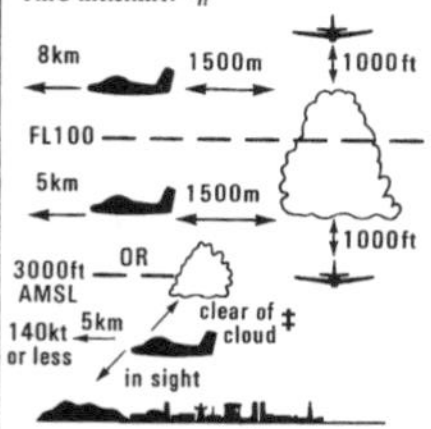

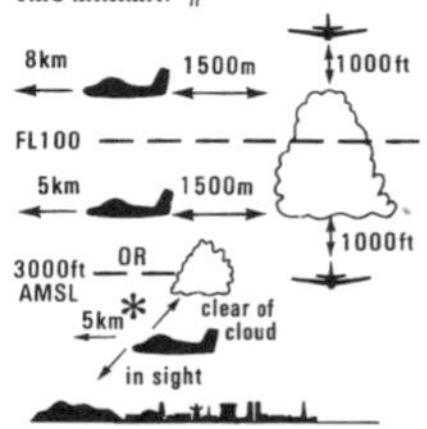

The Separation and Services provisions shown in the table are the minimum to meet ICAO Standards and Recommended Practices and may be supplemented when practicable. In particular, in Class F and Class G Airspace in the UK a Radar Advisory Service (RAS), a Radar Information Service (RIS) and Approach Control Service may be available from Air Traffic Service Units. Pilots are urged to make use of these services, details of which are published in AICs and other documents.

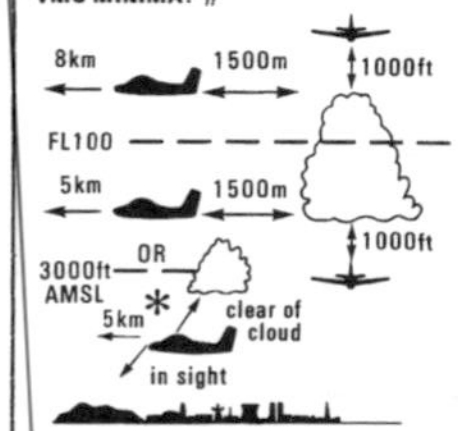

✳ At speeds of 140kt IAS or less flight is permitted in flight visibilities to 1500m. Helicopters may operate in less than 1500m flight visibility at a speed which, having regard to the visibility, is reasonable.

‡ Helicopters may fly at or below 3000ft AMSL clear of cloud and in sight of the surface.

ANO Schedule 8 & Licence Privileges may impose more stringent weather minima on Private Pilot's Licence and Basic Commercial Pilot's Licence holders.

+ Unless notified for ATC purposes.

Figure 7-1 UK VMC minima

UK CRITERIA FOR VFR FLIGHT

At and above FL100

Class B airspace	Visibility 8 km; clear of cloud
Class D, E, F & G airspace	Visibility 8 km, 1,500 m and 1,000 ft from cloud

Below FL100

Class D, E, F & G airspace	Visibility 5 km, 1,500 m and 1,000 ft from cloud

or

At or below 3,000 ft amsl

Class D and E airspace	Visibility 5 km, 1,500 m and 1,000 ft from cloud
■ at 140 kt or less	Visibility 5 km, clear of cloud and in sight of surface
■ helicopters	Clear of cloud and in sight of surface
Class F and G airspace	Visibility 5 km, clear of cloud and in sight of surface
■ at 140 kt or less	Visibility 1,500 m, clear of cloud and in sight of surface
■ helicopters	As above, but helicopters may operate in flight visibilities of less than 1,500 m at a speed which, with regard to visibility, is reasonable

■ *Figure 7-2* **Summary of VFR flight**

PRIVILEGES OF A PPL

In accordance with the Air Navigation Order, the privileges of a PPL permit the holder to operate within the following limitations:

Basic PPL

Minimum flight visibility 3 km

Remain in sight of surface at all times

Minimum visibility 10 km and in sight of surface on an SVFR clearance in a CTR

Flight in circumstances which require compliance with IFR not permitted

PPL with IMC Rating

Minimum flight visibility 1,500 m to maintain VFR

Minimum flight visibility 3 km, clear of cloud, and in sight of surface on an SVFR clearance in a CTR

Minimum visibility below cloud 1,800 m for any take-off or landing

Flight under VFR in Class B, D, E, F or G airspace in accordance with the VMC minima for the appropriate airspace shown in Figure 7-1

Flight in circumstances that require compliance with IFR not permitted in controlled airspace other than Class D and E

■ *Figure 7-3* **Summary of Private Pilot's Licence privileges**

Summaries of the more stringent limitations of the VFR minima and ANO Schedule 8, *Privileges for the PPL Holder in UK Airspace,* are shown in Figures 7-2 and 7-3.

The privileges of the Private Pilot's Licence are detailed further in Chapter 10.

VFR Flight Plan and ATC Clearance (Rule 27)

The pilot-in-command of a VFR flight must notify flight details to the appropriate Air Traffic Control unit in **Class** (B, C★ and) **D airspace**, and obtain an ATC clearance for the flight. (Responsible ATC units may, however, waive this requirement, at their discretion.)

The flight plan shall contain sufficient detail to enable the ATC unit to issue a clearance and for search and rescue purposes.

While operating in the specified airspace, the pilot-in-command shall maintain a constant listening watch on the appropriate radio frequency and comply with instructions from ATC.

Exemptions to this requirement are available in the specified airspace (effectively just Class D) for:

☐ gliders operating with a minimum flight visibility of 8 km, and distance from cloud of 1,500 metres horizontally and 1,000 ft vertically; and

☐ powered aircraft without radio equipment operating during the day with 5 km or better flight visibility and at least 1,500 metres horizontally and 1,000 ft vertically from cloud, provided the commander has previously obtained the permission of the appropriate ATC unit to make the flight.

Special VFR Flights (see also AIP ENR 1-2-2)

Clearance for a Special VFR (SVFR) flight is an authorisation by ATC to a pilot to fly within a Control Zone in marginal weather although being unable to comply with the Instrument Flight Rules.

Special VFR allows the relaxation by ATC, in certain circumstances, of some restrictions to facilitate the operation of a flight without lowering flight safety to an unacceptable level. SVFR is usually applied by ATC in **Class D** or **E Control Zones**, when weather and traffic conditions permit, to allow private pilots access to aerodromes within them, or to transit Class A airspace of a lower status, such as the London or Channel Islands CTRs. SVFR flights will not, however, be permitted outside Control Zones.

★ As well as Class D, the airspace specified in Rule 27 includes Classes B and C; however, Class B is only allocated to upper airspace (at and above FL245) and Class C is not currently allocated in the UK, so neither are of significance to usual VFR operations.

A flight plan is not required for an SVFR Flight, but ATC approval is. A request for a *Special VFR Clearance* may be made in flight, but will not necessarily be granted by ATC.

Authorisation for an SVFR flight is a *concession* granted by ATC only when weather and traffic conditions permit. An SVFR Clearance absolves the pilot from complying with:

- [ ] the full requirements of Instrument Flight Rules (IFR); and
- [ ] the requirement to maintain a height of 1,500 feet above the highest fixed object within 600 metres of the aircraft if the height limitation specified in the ATC clearance makes compliance with this requirement impossible. (It does *not* absolve the pilot from any other rule requirements, in particular the "ability to land clear in the event of engine failure" requirement of Rule 5, the low flying rule.)

All ATC instructions must be obeyed.

It is entirely the pilot's responsibility to ensure that flight conditions (i.e. forward visibility and distance from cloud) will enable the aircraft to remain clear of all obstructions. It is implicit in all SVFR clearances, therefore, that the aircraft remains clear of cloud and in sight of the surface. Flight visibility must be at least 10 km (unless the pilot holds an IMC Rating, when the minimum flight visibility required is reduced to 3 km).

At some aerodromes in Control Zones there are designated **Entry/Exit Lanes** in which aircraft may fly in Instrument Meteorological Conditions (IMC) without complying with the full Instrument Flight Rules procedures, and in most cases without having to obtain clearance. The procedures and weather minima for flying in these lanes are given in the UK Aeronautical Information Publication (AIP AD 2-22) and in *Pooley's Flight Guide.* This is also covered in more detail in Chapter 10, *The Private Pilot's Licence.*

Flights at Night

Night is defined in the Air Navigation Order as meaning the time between 30 minutes after sunset and 30 minutes before sunrise, as measured at surface level.

Flights under the Visual Flight Rules at night are not permitted in the UK. However, flights at night in VMC *are* permitted but must be conducted in compliance with IFR Rule 29 (minimum height) and IFR Rule 30 (cruising levels above 3,000 ft or the transition altitude). These two rules are covered in the next chapter: Instrument Flight Rules.

Notes on the Visual Flight Rules

In interpreting the Visual Flight Rules it is essential to remember that a pilot must at all times fly within the privileges of the licence which is held.

In the interest of safety, pilots are **advised** to select cruising levels in accordance with the *quadrantal* or *semicircular* rule as applicable, even when under VFR. However, it is perfectly acceptable, when under VFR, to fly on the regional QNH pressure setting above 3,000 ft, especially when manoeuvring.

Now complete **Exercises 7 – Visual Flight Rules (Rules 24–27).**

Instrument Flight Rules (Rules 28–32)

Flight under Visual Flight Rules (VFR) is very restrictive if, for instance, a regular air service is to be achieved. For this reason, flight and navigation instruments have been developed that allow a properly trained pilot to operate in cloud and other conditions not suitable for visual flight. The rules that apply to this category of flight are known as the **Instrument Flight Rules (IFR)**.

Only two types of flight category are available in aviation: VFR or IFR.

The Minimum Height Rule (Rule 29)

Subject to the usual low flying requirements (Rule 5), an aircraft operated according to the Instrument Flight Rules, both inside and outside controlled airspace, must not fly at less than 1,000 feet above the highest obstacle within 5 nm, except:

1. On a route notified for the purposes of this rule or otherwise authorised by the competent authority (this may include controlled airspace such as certain Terminal Control Areas and Airways with particular high obstacles such as a radio mast underlying them and giving less than 1000 ft clearance from the base of the controlled airspace);

2. As necessary for take-off or landing; and

3. When flying at 3,000 feet amsl or below, clear of cloud and in sight of the surface.

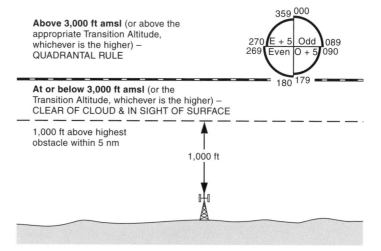

■ *Figure 8-1* **General IFR requirements**

IFR Flights Cruise at Quadrantal Levels (Rule 30)

In level flight above 3,000 ft amsl (or above the appropriate transition altitude, whichever is the higher), pilots must select cruising levels according to the *quadrantal rule* (or *semicircular rule* above FL245). *Flight levels* are based on the standard altimeter setting 1013.2 mb (or hPa).

The quadrantal and semicircular rules require an aircraft to be flown at a cruising level appropriate to its **magnetic track**, as shown below.

IFR FLIGHTS AT LEVELS BELOW FL245		
Magnetic Track	*Cruising Level*	
000°–089°	*Odd thousands of feet*	
090°–179°	*Odd thousands plus 500 feet*	
180°–269°	*Even thousands of feet*	
270°–359°	*Even thousands plus 500 ft*	

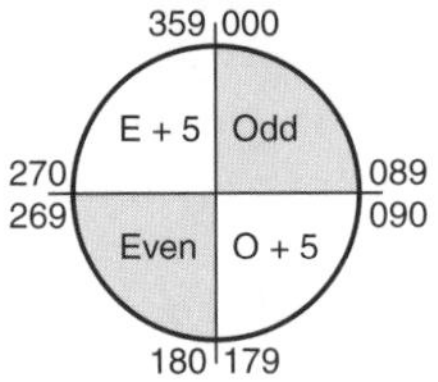

IFR FLIGHTS AT LEVELS ABOVE FL245	
Magnetic Track	*Cruising Level*
000°–179°	*FL250 (lowest usable)*
	FL270
	FL290 and higher levels at intervals of 4,000 feet
180°–359°	*FL260 (lowest usable)*
	FL280
	FL310 and higher levels at intervals of 4,000 feet

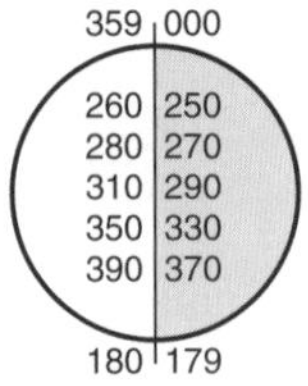

Figure 8-2 **IFR cruising levels for below and above FL245**

An aeroplane operating under the Instrument Flight Rules may cruise at a different level from those specified above when complying with instructions given by an ATC unit or with notified holding procedures.

Flight Plan and ATC Clearance (Rule 31)

Rule 31 applies to IFR flights within controlled airspace.

1. In order to comply with the Instrument Flight Rules, before any flight within controlled airspace the aircraft commander must file a flight plan (irrespective of whether IMC or VMC exist) and obtain an ATC clearance based upon it. The flight must be made in accordance with the clearance and with the notified holding and approach procedures at the destination unless otherwise instructed by ATC.

2. A pilot flying IFR in controlled airspace must follow:

- the terms of the Air Traffic Control clearance and any further instructions given by ATC;
- the published instrument holding and approach procedures for the destination aerodrome;
  however, he may cancel IFR (and therefore switch to VFR), provided that:
  > he can maintain VMC whilst in controlled airspace; and
  > he informs ATC accordingly, asking them to cancel his flight plan.

3. ATC must be told as soon as possible if, to avoid immediate danger, any departure has to be made from the requirements of this rule.

4. Except when the flight plan has been cancelled, an aircraft commander must inform ATC when the aircraft lands within or leaves controlled airspace.

Position Reports (Rule 32)

An aircraft under IFR which flies in or intends to enter controlled airspace must report its time, position and level at such reporting points or at such intervals of time as may be notified or directed by ATC.

TYPICAL POSITION REPORT	
Aircraft identification	Golf Alpha Echo Sierra Echo
Position and time	Wicken four seven
Level	Flight level four zero
Next position and estimate	Marlow five seven

Applicability (Rule 28)

1. Rules 29, 31 and 32 apply to flight operations within controlled airspace.

2. Rules 29 and 30 apply to flight operations outside controlled airspace.

Now complete
Exercises 8 – Instrument Flight Rules (Rules 28–32).

Registration and Airworthiness

Registration and Marking of Aircraft

ICAO Annex 7

Aircraft Nationality and Registration Marks

Annex 7 contains standards and definitions adopted by ICAO to define the minimum requirements for the display of aircraft nationality and registration marks.

Nationality and Registration Marks to be Used

2.1 The nationality and registration mark shall consist of a group of letters and/or numbers (e.g. G-AGOH, N7207V).

2.2 The nationality mark shall appear before the registration mark in the group. When the first character of the registration mark is a letter it shall be preceded by a hyphen (e.g. F-BUDG).

2.5 The registration mark shall be letters and/or numbers, and shall be assigned by the State of Registry.

Location of Nationality and Registration Marks

3.1 The nationality and registration mark shall be painted on, or permanently affixed to, the aircraft. The marks shall be kept clean and visible at all times.

3.2 On balloons, the marks shall appear on each side, near the widest part of the balloon, above either the rigging band or the basket cable attachment points. On other lighter-than-air aircraft, the marks shall be visible from the sides and from the ground.

3.3 On heavier-than-air aircraft the marks shall appear once on the lower surface of the wing, on the left-hand side unless they extend across the whole wing structure. The tops of the letters and/or numbers shall be toward the leading edge of the wing, and the mark shall be positioned midway between the leading and trailing edge.

The marks shall also appear on both sides of the fuselage, between the wings and tail surface, or on the upper halves of the vertical tail surfaces. When located on a single vertical tail surface the marks shall appear on both sides.

Dimensions of Nationality and Registration Marks

4.1 On lighter-than-air aircraft, the height of the marks shall be at least 50 centimetres.

4.2 For heavier-than-air aircraft, the height of the marks on the fuselage shall be at least 50 centimetres. On the fuselage or vertical tail surfaces the marks shall be at least 30 centimetres high.

Lettering for Nationality and Registration Marks

5.1 Letters shall be Roman capitals, without ornamentation. Numbers shall be Arabic, without ornamentation. The width of each character (except for the letter I and the number 1) shall be two-thirds of the height of a character. In other words, the character shapes cannot be 'stretched' too much in height as to make them difficult to read. The characters should be formed from solid lines, in a contrasting colour to the background.

Identification Plate

Aircraft shall carry an identification plate inscribed with its nationality or common mark and registration mark. The plate shall be made of fireproof metal or other suitable fireproof material, and shall be secured to the aircraft near the main entrance.

Registration of Aircraft in the UK
(ANO Articles 3, 4 and 5; Schedule 2)

To all intents and purposes, the private pilot must not fly in the UK unless the aircraft is registered, either in the UK or elsewhere, and displays the appropriate registration markings.

■ *Figure 9-1* **Part of a Certificate of Registration**

■ *Figure 9-2* **Registration markings**

In applying for registration, the aeroplane must be properly described according to the General Classification of Aircraft specified in Schedule 2 of the Air Navigation Order (ANO).

It is the responsibility of **owners** (and part-owners) to inform the CAA in writing of any change in the particulars shown on the original application for registration of an aircraft, e.g. change of ownership or part-ownership, its destruction or permanent withdrawal from use.

Gliders and hang-gliders need not be registered. In some circumstances, a powered aeroplane need not be registered, e.g. experimental or test aeroplanes, and aeroplanes undergoing certification testing, or carrying out a demonstration flight.

If a pilot does fly an unregistered aircraft, he is accountable for any other offences against the ANO made during the flight, just as if the aircraft were registered.

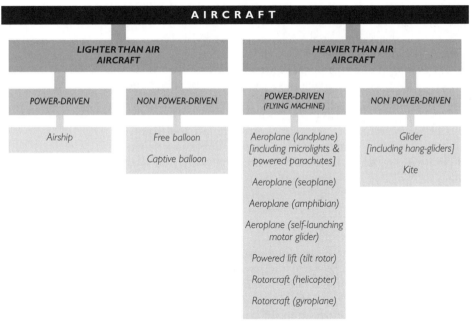

■ Figure 9-3 **The general classification of aircraft**

Airworthiness and Equipment

Airworthiness of Aircraft (ICAO Annex 8)

Annex 8 describes the broad standards of airworthiness to be adopted by national authorities. These standards are the minimum basis by which a Certificate of Airworthiness issued by one state can be recognised by another State. This ensures that when aircraft fly over the territory of other States, they comply with international minimum airworthiness standards, which are agreed by all contracting States. This ensures a universal standard of safety and protection to aircraft, people and property in all countries.

Airworthiness of Aircraft (ANO Articles 8–12, Schedule 3)

These ICAO standards of airworthiness are not intended to replace national regulations or codes of practice, due to the differences in detail considered necessary by individual States.

In the UK, legislation on airworthiness of aircraft is contained in the Air Navigation Order (ANO) and other CAA documents. Fine engineering and maintenance details are described in the British Civil Aviation Regulations (BCARs) – large volumes which are essentially very detailed technical manuals. However, in the light of European harmonisation (standardisation of regulations between countries), very close airworthiness and maintenance standards are being agreed with our European partners. These are described in the Joint Aviation Requirements (JARs), which embrace the requirements of ICAO. Now that the European Aviation Safety Agency has come into being, the JARs will be subsumed into a new format during
the next few years.

A Certificate of Airworthiness (CofA) is issued by the CAA in respect of each aeroplane determined to be airworthy, following consideration of design, construction, workmanship, materials, essential equipment and the results of flying trials and other tests. In the UK a CofA usually lasts for 3 years and will have a validity period specified on it. Certification, especially of a new aeroplane type, is a very important process and a major function of the CAA.

NOTE The Certificate of Airworthiness is considered invalid if the aircraft is modified, repaired or maintained in other than an approved manner (this in turn would invalidate the aircraft's insurance).

A **Flight Manual** will form part of the CofA and will specify requirements, procedures and limitations with respect to the operation of the aeroplane. In the case of foreign-manufactured aircraft, the CAA may attach a supplement to the original Flight

Manual, which amends the manner in which the aeroplane should be operated.

In general, a private pilot may only fly within the UK in an aeroplane that has a **current CofA** valid under the law of the country in which it is registered and then only **according to the conditions specified;** e.g. to tow a glider legally, the CofA must be so endorsed (Article 54). The CofA will specify the **categories** (transport, aerial work, private, or special) in which the aeroplane may be operated (ANO Article 9, Schedule 3).

In some circumstances, an aeroplane that does not have a current CofA may be issued with a *Permit to Fly* by the CAA, subject to certain conditions. (See 'A and B Conditions' in ANO Schedule 3.) This could apply to an aeroplane undergoing tests, but also applies to the majority of 'home-built' aeroplanes. In the latter case, the Permit may be issued by the Popular Flying Association (PFA).

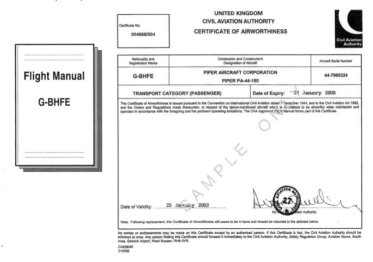

■ Figure 9-4 **The Flight Manual forms part of the CofA**

Maintenance (ANO Articles 10, 11, 12)

If an aircraft has a CofA in either the *transport* or *aerial work* category, then the aircraft and its equipment must be maintained to a schedule approved by the CAA and it must have a valid **Certificate of Maintenance Review.**

A **Technical Log** must also be kept for such aircraft to record aircraft flight times, unserviceable items, maintenance carried out, etc. Technical defects which occur should be entered in the technical log before the aircraft is flown by another pilot or at the end of the day, whichever is the earlier.

NOTE *Flying training* is classified as aerial work and hired aero-planes fall into the *transport* category, so this requirement applies to aeroplanes that a private pilot is likely to fly.

An aircraft must have, in addition to a CofA, a **Certificate of Release to Service** if it (or any part of it or such of its equipment as is necessary for its airworthiness) has been overhauled, repaired, replaced, modified or otherwise maintained. Only certain persons are allowed by the ANO to issue such a certificate, although the Air Navigation (General) Requirements (ANGRs) specify minor repairs or replacements which may be made by a pilot to aircraft in the Private or Special categories that are less than 2,730 kg maximum weight, such as replacing a battery, changing tyres, completing an oil change, changing spark plugs and setting their gaps, replacing a VHF-COM set (provided it has no connection with navigation equipment), etc. For more details see ANGR Reg. 16.

Provision is also made for the pilot of an aeroplane that is **away from base** to confirm that a minor adjustment to a control has been satisfactorily performed and to sign the second part of a Duplicate Inspection Certificate.

Engine and Propeller Logbooks (Article 17 and Schedule 6). In addition to the previously mentioned documents, the *operator* is required to keep:

- an Aircraft Logbook;
- an Engine Logbook;
- a Variable-Pitch Propeller Logbook (if applicable).

All logbooks have to be preserved for two years after the particular aircraft, engine or propeller has been destroyed or permanently withdrawn from use.

Temporary Loss of Airworthiness (ICAO Annex 8)

PART II, 6.1 If an aircraft is not maintained in an airworthy condition as defined by the appropriate requirements it cannot be flown until it is restored to an airworthy condition.

PART II, 6.2 If an aircraft is damaged, the State of Registry shall determine whether such damage renders the aircraft no longer airworthy.

If an aircraft is damaged in the territory of another State, that State is entitled to prevent the aircraft from flying, provided they notify the State of Registry immediately. Should the State of Registry consider that the damage sustained makes the aircraft not airworthy, it shall prohibit the aircraft from flying until it is restored to airworthiness. However, in exceptional circumstances it may allow the aircraft to fly under limiting conditions, without fare-paying passengers, to an aerodrome where repairs can be made. The State in which the aircraft was damaged (that prevented further flight in the first place) shall allow such a flight to be made.

On the other hand, if the State of Registry considers that the damage sustained (in another State) is such that the aircraft is still airworthy, the aircraft shall be allowed to resume its flight.

Weight and Balance (ANO Article 18)

An important part of the Certificate of Airworthiness is the **Aircraft Weight Schedule**, which approves certain weight limitations and specifies the allowable range for the position of the centre of gravity (CG).

ANO Article 18 requires that an aeroplane must be weighed at its basic weight (or other approved weight) and have the position of its centre of gravity determined at that weight. A Weight Schedule must then be prepared by the operator. Following the next occasion when the aircraft is officially weighed for the purposes of this Article and a new Weight Schedule prepared, the earlier Weight Schedule should be preserved for a period of 6 months, even though it has expired.

Fly within the CofA Specifications

Operating the aircraft outside the conditions specified in the CofA (which includes the Flight Manual) may seriously degrade the safety of the flight and render the CofA invalid, e.g. flying a non-permitted operation or manoeuvre or exceeding *weight and balance* limitations. In addition to the safety issues, this has very serious legal implications, making the pilot liable to a fine or imprisonment, rendering the insurance policies invalid and invalidating any warranties or guarantees on the aircraft and its equipment.

A pilot should always operate an aeroplane within the specifications of its CofA.

Aircraft Limitations and Information (ICAO Annex 8)

PART II, 8 Each aircraft shall be provided with a Flight Manual, placards, or other documents describing any limitations within which the aircraft is considered airworthy, and any other information necessary for the safe operation of the aircraft.

Operating Limitations and Information

PART III, 9.1 Any operating limitations which must be complied with under the Certificate of Airworthiness, and any other information necessary for the safe operation of the aircraft, shall be made available in a Flight Manual or on placards or markings etc.

PART III, 9.2 Where there is a risk of exceeding operating limitations in flight, they shall be defined such that the flight crew can determine, by reference to the instruments available to them, when the limitations are reached.

Loading limitations shall include all limiting mass (weights), centres of gravity positions, mass (weight) distributions and floor loadings.

Airspeed limitations shall include all speeds that are limiting in relation to structural integrity and flying qualities (e.g. $\hat{V}_{NE}$). Various aeroplane configurations shall be taken into account in identifying these speeds (e.g. V_{FE}).

Powerplant limitations shall include all those established for the various powerplant components installed in the aeroplane.

Equipment and systems limitations shall include all those established for the various equipment and systems installed in the aeroplane.

Miscellaneous limitations shall include any other limitations necessary to the safe operation of the aeroplane.

PART III, 9.3 Loading information shall include the empty mass of the aeroplane, together with a definition of the condition of the aeroplane at the time of weighing, the corresponding centre of gravity position, and the reference point(s) and datum line(s) to which the CG limits are related.

NOTE 'Empty mass' usually excludes crew and payload, usable fuel and drainable oil; but it includes all fixed ballast, unusable fuel, undrainable oil, engine coolant and hydraulic fluid.

Operating procedures, both normal and emergency, that are unique to the aircraft shall be described. These shall include engine-failure procedures.

Handling information. Unusual or significant aeroplane characteristics, including stall speeds or minimum steady flight speeds, shall be described.

PART III, 9.5 An aeroplane Flight Manual shall be available to the crew. It shall identify clearly the specific aeroplane or series of aeroplanes for which it is written and contain at least the limitations, procedures and information mentioned above.

PART III, 9.6 Markings and placards on instruments, equipment, controls, etc., shall include any limitations or information necessary to the crew during flight.

Electrical Systems

PART IV, 8 Aircraft electrical systems shall be designed and installed to ensure that they will perform their intended function under any foreseeable operating conditions.

Equipment of Aircraft
(ANO Articles 14, 15 & 16, Schedules 4 & 5)

Many aircraft are lavishly equipped, but there is a certain minimum standard specified in the ANO below which an aeroplane should not be flown (general aircraft equipment being detailed in ANO Schedule 4, and radio communication and radio navigation equipment being detailed in Schedule 5).

Pilots must be familiar with the minimum equipment requirements for their particular operation; e.g. when carrying out aerobatic manoeuvres, a suitable safety harness or safety belt must be fitted; when night flying in helicopters, parachute flares must be carried; an unserviceable attitude indicator (artificial horizon) may not render the aeroplane unserviceable, whereas an unserviceable clock may do so.

Noise Certification

A Noise Certificate is required for every propeller-driven aeroplane with an all-up weight of 9,000 kg or less, all helicopters and microlight aeroplanes, and every other subsonic aeroplane which at maximum authorised weight requires a take-off distance of more than 610 metres. Certain short take-off and landing (STOL) aeroplanes are exempt.

Insurance

While there is no legal requirement in the UK to insure an aircraft, flying an uninsured aircraft could lead to you being substantially out of pocket, even bankrupt, if you suffer an accident.

As part of your pre-flight check you should ensure that the certificate of insurance is present and valid for the type of flight intended. Although it is the responsibility of the owner or operator to arrange for insurance cover, you should remember that you, as pilot-in-command, are responsible under Article 43 of the ANO for ensuring that the flight may be safely and legally conducted. The existence of insurance cover does not exempt you from observing the requirements of the ANO and your licence privileges.

If you hire an aircraft from an FTO or RF it is likely that proper and adequate insurance will be in place, although beware, there may be an 'excess' that you would have to pay in the event of an accident. The excess is likely to be between £500 and £1,000 depending on the type of aircraft, however some clubs operate a scheme to self-insure the excess by contributing a small amount to a pooled kitty (which obviously will have to be replenished by further contributions if it has to be drawn on following someone else's accident).

If you wish to land at a military airfield you may have to arrange additional third party cover up to £7,500,000 or pay a premium on top of the landing fee. This is because at present typical light aviation insurance policies only cover £1,000,000 of third party liability and all that military hardware is rather expensive if you accidentally run into it on landing! This limit is likely to be raised considerably by new European legislation requiring all aircraft to carry higher minimum insurance cover.

You may find a number of exclusions in your insurance policy, so do check it carefully. Typically your insurer may not cover you for landing at unlicensed strips, or perhaps for aerobatics. You may also find that there is a 'pilot warranty' clause requiring a minimum number of hours total time or experience on type. Be careful of this as insurers are well-known for refusing to pay out on claims arising from accidents where the pilot has fewer hours than specified in the pilot warranty. If you are not the owner you could then face a claim personally by the owner, in addition to any third party claims! If in doubt, check with your insurance broker.

Now complete **Exercises 9 – Registration and Airworthiness.**

Pilots' Licences

JAR-FCL

Annex 1 of the Chicago Convention refers to the licensing requirements and recommendations of the ICAO signatories. Approximately half of the 176 member States have filed differences for their national licensing requirements. The JAA's Flight Crew Licensing Committees worked for a number of years to produce the document now known as JAR-FCL1 (Joint Aviation Requirements Flight Crew Licensing [Aeroplane]). The aim was to harmonise flight crew licensing throughout Europe in order to simplify the movement of pilots from one country to another. This clearly has benefits for commercial pilots and those wishing to earn a living from flying.

The UK was the first State to adopt the new requirements, which were implemented with effect from July 1999. The JAR-FCL documents cover flying training and testing of pilot licences and ratings for all member States of the JAA. JAR-FCL2 covers helicopters and JAR-FCL3 covers medical certification.

The use of JAR-FCL licences without further written formality in all participating States represents a significant development for international aviation. However, the changes are complex and their implementation into UK law is leading to a continual process of fine-tuning as the system takes effect. With the introduction of EASA, more changes are likely. Check with your FTO and flight instructor for further developments.

The UK CAA PPL

Revalidation is the administrative action which takes place in order to revalidate a licence or rating before the privileges have lapsed.

Holders of an existing UK CAA PPL will not need to convert this to a JAA PPL, but they can only exercise the full privileges of the UK CAA PPL on UK-registered aircraft. Before exercising their licence privileges in aircraft registered in other member States, they may have to comply with additional requirements of those other States. As training for the UK CAA PPL is no longer available, existing holders of a UK CAA PPL will have to revalidate their licences in accordance with JAR-FCL. However, this action is simply for the revalidation of ratings. It does not accord holders the privileges of a JAR-FCL licence.

JAR-FCL 1-3

The changes introduced by JAR-FCL have been incorporated into UK law through the ANO 2000, which took effect from July 2000. There are a number of new definitions with which you will need to become familiar. These are summarised on the following pages.

JAR-FCL TERMINOLOGY

Category (of aircraft)
A grouping of similar aircraft according to specified basic characteristics, e.g. aeroplane, helicopter, glider, free balloon.

Class rating
Approval to fly a single-pilot aeroplane that does not require a type rating.

Conversion (of a licence)
The issue of a JAR-FCL licence on the basis of a licence issued by a non-JAA State.

Dual instruction time
Flight time or instrument ground time during which a person is receiving flight instruction from a properly authorised instructor.

Flight time
The total time from the moment that an aircraft first moves under its own or external power for the purpose of taking off until the moment it comes to rest at the end of the flight.

Flight time as student pilot-in-command (SPIC)
Flight time during which the flight instructor will only observe the student acting as pilot-in-command and shall not influence or control the flight of the aircraft.

Instrument time
Instrument flight time or instrument ground time (i.e. attained in a simulator).

Instrument flight time
Time during which the pilot is controlling an aircraft solely by reference to the flight instruments.

Instrument ground time
Time during which a pilot is receiving instruction in simulated instrument flight in synthetic training devices (STDs).

Multi-crew cooperation
The functioning of flight crew as a team of cooperating members led by the pilot-in-command.

Multi-pilot aeroplanes
Aeroplanes certificated for operation with a minimum crew of at least two pilots.

Night
The period between the end of evening civil twilight and the beginning of morning civil twilight, or such other period between sunset and sunrise as may be prescribed by the appropriate Authority.
Note: Night is defined in the UK as 30 minutes after sunset until 30 minutes before sunrise.

Other training devices
Training aids other than flight simulators; flight training devices or flight and navigation procedures trainers that allow training where a complete flight deck environment is not necessary.

Private pilot
A pilot who holds a licence that does not permit him to be paid for flying aircraft.

Professional pilot
A pilot who holds a licence that permits him to be paid for flying aircraft.

JAR-FCL TERMINOLOGY

Proficiency checks
Demonstrations of skill to revalidate or renew ratings, including oral examinations as necessary.

Rating
An entry in a licence stating special conditions, privileges or limitations pertaining to that licence.

Renewal
*(e.g. of an **expired** rating or approval) The reinstatement of the privileges of the rating or approval for a further specified period, subject to certain requirements.*

Revalidation
*(e.g. of an **existing** rating or approval) Permission granted that allows the holder of a rating or approval to exercise the privileges of the rating or approval for a further specified period, subject to certain requirements.*

Single-pilot aeroplanes
Aeroplanes certificated for opera-tion by one pilot.

Skill tests
Demonstrations of skill for the issue of a licence or rating, including oral examinations as required.

Solo flight time
Flight time during which a student pilot is the sole occupant of an aircraft.

Touring motor glider (TMG)
A motor glider with a Certificate of Airworthiness issued or accepted by a JAA member State that has an integrally mounted, non-retractable engine and a non-retractable propeller. It shall be capable of taking off and climbing under its own power according to its Flight Manual.

Type (of aircraft)
Aircraft of the same basic design, including all modifications except those that result in a change of handling, flight characteristics or flight crew complement.

Type rating
*A rating that permits the holder to fly a specific aeroplane type. A type rating is **not** included in a class rating.*

Student Pilots

A student pilot may commence training without a licence under the supervision of a suitably qualified flight instructor. During the course of training for a JAR-FCL PPL(A), a student pilot may (without a licence) act as pilot-in-command (PIC) and fly solo if the following conditions are complied with.

- The student pilot is at least 16 years of age.
- The student pilot holds a medical certificate validated by an approved Authorised Medical Examiner (AME) and complies with any restrictions stated thereon (e.g. a requirement to wear spectacles if decided by the AME).

☐ No other person is carried in the aircraft.

☐ The student pilot acts in accordance with instructions given by a suitably qualified flight instructor.

Validity of Medical Certificates

Before applying for a licence and before exercising the privileges of a licence, the applicant or licence holder must have passed a medical examination made by an Authorised Medical Examiner (AME).

The periods of validity for medical certificates issued on or after 1 July 1999 are as follows:

MEDICAL CERTIFICATE VALIDITY – CLASS 2	
Age	Validity
less than 30 years	60 months
30 to 49 years	24 months
50 years or over	12 months

Note 1 This certificate is valid for its time period even if the applicant enters the next age group. The only exception is that if a JAA Class 2 certificate is issued prior to the 30th birthday (validity period normally five years), it is not valid beyond the pilot's 32nd birthday. Certificates issued prior to July 1999 remain subject to the old rules and the calculated 'date of next examination' validity periods remain as stated on the certificate. Your AME will be able to answer queries on this issue.

Note 2 Professional pilots are required to hold a Class 1 medical certificate, which is more stringent.

Note 3 Pilots may renew their medical certificates up to 45 days before the expiry date. The validity period will commence from the expiry date of the previous medical certificate.

NOTE An alternative form of medical certificate is available for flight of balloons and airships (or microlight aircraft) only. It has a **declaration of health** to be signed by the applicant and counter-signed by his or her general practitioner, or an AME. The certificate is only valid if so countersigned. Its period of validity is the same as for the standard certificate. Additionally the medical standards for the National Private Pilot's Licence (NPPL) are lower. See page 136 for more details.

In all cases, it is the licence-holder's responsibility to ensure that renewals are made within the appropriate time limits if licence privileges are to be maintained.

A pilot is not entitled to act as flight crew if he knows or suspects that he is **medically unfit** for it. He must therefore seek advice from an AME or the CAA Medical Department immediately. Conditions include:

☐ admission to a hospital or clinic for more than 12 hours;

☐ undergoing a surgical operation or invasive medical investigation;

Renewal is the administrative action which takes place in order to revalidate a licence or rating after the privileges have lapsed.

☐ frequent use of medication;

☐ requiring to use corrective lenses on a regular basis.

ANAESTHETICS A pilot may not exercise the privileges of his licence after receiving a local anaesthetic for at least 12 hours after the event (such as a dental visit requiring an injection for tooth extraction or filling), and at least 24 hours, but possibly up to 72 hours after a general anaesthetic (see AIC 143/1998 (Pink 183)).

BLOOD OR BONE MARROW DONATION Twenty-four hours should elapse between giving blood and flying. In the case of bone marrow donation involving a general anaesthetic, 48 hours should elapse before flying (see AIC 14/1998 (Pink 166)).

PREGNANCY, ILLNESS OR INJURY The CAA must be advised in writing as soon as possible in the case of injury or pregnancy, and as soon as the period of 21 days has elapsed in the case of illness. The medical certificate is deemed to be suspended from the date of the occurrence until such time as the individual is re-examined and certified medically fit again.

The Granting of a JAR Private Pilot's Licence (PPL)

The CAA only grants a PPL when it is satisfied that the applicant is a fit person to hold the licence and is qualified by reason of knowledge, experience, competence, skill, physical fitness and mental fitness. The minimum age for holding a Private Pilot's Licence is 17 years.

In addition to holding a valid medical certificate, a PPL applicant must produce evidence to the CAA that he has, within an 18-month period from the date of the first examination, and within a period not more than two years preceding the granting of the licence, passed:

☐ **seven ground examinations**: (1) Aviation Law, Flight Rules and Operational Procedures); (2) Navigation; (3) Meteorology; (4) Aircraft (General); (5) Human Performance and Limitations; (6) Flight Performance and Planning and (7) Communications – (plus a practical examination in Radiotelephony);

☐ **a solo navigational cross-country flight** of at least 150 nm with two intermediate full stop landings at aerodromes other than the aerodrome of departure;

☐ **a flight test,** which will also involve an oral examination: Aircraft (Type).

The JAR Private Pilot's Licence is not valid until it is signed in ink by the holder. It is then valid for 5 years from the date of issue.

UK PPL VALIDITY PERIOD The old UK PPL has no maximum validity period, although, for the holder to exercise its privileges, it must be renewed periodically in the same way as a JAR PPL with a **Certificate of Revalidation**, which is valid for 24 months.

JAR PPL RATINGS It is possible to obtain a JAR PPL from any member State and then obtain ratings in another member State, although the ratings will have to be entered on the licence by the State of licence issue. If the licence holder wishes to transfer the licence to another state for administrative convenience, he will have to show 'normal residency'. (This means the place where he resides and/or works for more than 185 days each year.)

The Privileges Accorded to a Private Pilot

A Private Pilot's Licence holder may fly as *pilot-in-command* (PIC) (often stated as *'aircraft commander')* or as *co-pilot* in any of the aeroplane types or classes specified in the aircraft rating in the licence, provided:

- ☐ He has included in his pilot's licence a Certificate of Revalidation (CofR) valid for the type or class of aircraft to be flown.

- ☐ He has a valid medical certificate issued by an AME.

- ☐ He does not fly an aeroplane for the purpose of public transport (i.e. the carriage of persons or cargo for hire or reward, either promised or given) or aerial work (i.e. flight for hire and reward for purposes not involving public transport) other than aerial work which consists of:
  - the giving of flight instruction, in which case his licence must be endorsed with a Flight Instructor's Rating or a Restricted Flight Instructor's Rating; or
  - the conducting of tests for the purposes of the ANO.

NOTE In either case the aerial work would have to be done in an aircraft owned, or operated under arrangements entered into, by a flying club of which the person instructing or conducting the test and the person being instructed or undergoing the test are both members.

- ☐ He may fly an aeroplane for the purpose of aerial work which consists of:
  - towing a glider in flight; or
  - a flight for the purpose of dropping persons by parachute.

NOTE Similar to the previous note, in either of the above cases the aeroplane must be owned, or operated under arrangements entered into, by a club of which the licence holder and any person carried in the aircraft or in any glider towed by the aircraft are members.

- ☐ He is not remunerated for services as a pilot other than, if his licence includes a Flight Instructor's Rating or a Restricted Flight Instructor's Rating by virtue of which he is entitled to give instruction in flying microlight aircraft or self-launching motor gliders, he may receive remuneration for instruction in such aircraft.

NOTE Cost-sharing between pilot and passengers on private flights is permitted, providing the conditions in ANO Article 130 are met. The essential points are that no more than four persons are carried and the pilot's share of the costs must be at least a full share based on the number of persons carried. The flight cannot be advertised and the pilot must not be employed by the aircraft operator.

▢ Unless he has obtained (and his licence includes) an Instrument Rating (IR) or Instrument Meteorological Conditions (IMC) Rating, he does not fly as pilot-in-command:
  (i) when outside controlled airspace:
    > in a flight visibility (i.e. visibility forward from the cockpit) of less than 3 km and out of sight of the surface.
  (ii) on a Special VFR (SVFR) flight in a control zone in a flight visibility of less than 10 km except on a route or in an Aerodrome Traffic Zone (ATZ) notified for the purpose;
  (iii) out of sight of the surface;
  (iv) or co-pilot, of an aeroplane in Class D or E airspace in circumstances that require compliance with the Instrument Flight Rules (IFR).

▢ He does not fly as pilot-in-command at night (which is defined in the ANO as from 30 minutes after sunset to 30 minutes before sunrise), unless:
  (i) his licence includes an aeroplane Night Rating or Night Qualification; and
  (ii) his licence includes an Instrument Rating (IR) or he has, within the immediately preceding 90 days, completed at least three take-offs and landings. If passengers are to be carried at night, one take-off and one landing within the preceding 90 days must have been made at night.

NOTE A PPL holder (without a Night Rating, Night Qualification or an Instrument Rating) may also fly as pilot-in-command (PIC) at night in accordance with the instructions of an authorised flight instructor and provided no passengers are carried.

▢ He does not, unless his licence includes an Instrument Rating, fly as pilot-in-command or co-pilot of an aeroplane in Class A, B or C airspace in circumstances that require compliance with the Instrument Flight Rules.

Carriage of Passengers – Recency Requirements

In order to carry passengers the pilot-in-command must have made at least three take-offs and landings within the preceding 90 days, as the sole manipulator of the controls of the same type or class of aircraft to be flown. If passengers are to be carried at night, one out of the three take-offs and landings within the preceding 90 days must have been made at night.

Flying VFR on a Private Pilot's Licence

Figure 10-1 summarises the VFR minima for PPL holders, by combining the more stringent requirements of:

☐ the Visual Meteorological Conditions (see Figure 7-1); and

☐ the licence privileges, including the need to observe the requirements for operations within controlled airspace available (i.e. not Class A).

Weather Minima for VFR Flight in the United Kingdom
by PPL Holders – no IMC or Instrument Rating (Below FL245)

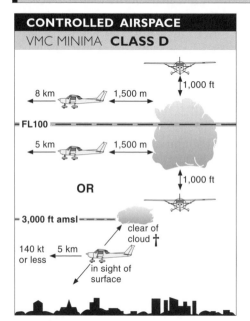
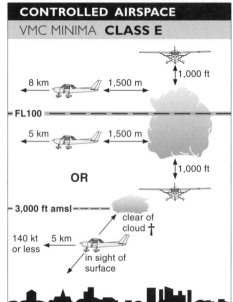

† Helicopters may fly at or below 3,000 feet amsl clear of cloud and in sight of the surface.

Note Special VFR is permitted, but minimum visibility is 10 kilometres clear of cloud and in sight of the surface (except on Notified Routes or in Notified ATZ).

■ Figure 10-1 **VFR weather minima**

NOTE If you do not hold a particular licence or rating, you cannot undertake any of the functions to which that licence or rating relates.

Weather Minima for VFR Flight in the United Kingdom
by PPL holders – no IMC or Instrument Rating (Below FL245)

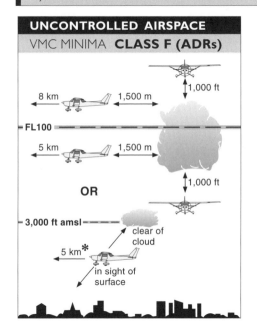

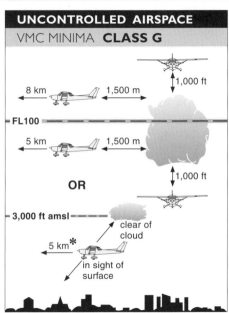

* *For non-rated IMC or IR pilots, at or below 3,000 feet amsl at speeds of 140 knots IAS or less, VFR flight is permitted in flight visibility as low as 3 kilometres.*
Helicopters may operate in less than 1,500 metres flight visibility at a speed which, having regard to the visibility, is considered reasonable.

■ *Figure 10-2* **VFR weather minima (continued)**

The Flight Radiotelephony Operator's Licence

To operate radio-equipped aircraft, a private pilot should hold a **Flight Radiotelephony Operator's (FRTO) Licence** (although student pilots have a dispensation). A written and practical classroom test is required to attain the licence; your training organisation will guide you on this. The licence is a stand-alone licence, and although the applicant for a PPL has to have passed the Communications paper prior to applying for his pilot's licence, he does not have to have completed the practical test. Refer also to Volume 7 of *The Air Pilot's Manual, Radiotelephony*, for more details.

PPL (Helicopters and Gyroplanes)

While the requirements relating to PPL licences for helicopters and gyroplanes mirror closely the requirements for the PPL(A), some differences should be noted.

The holder of UK PPL (Gyroplanes) must not:

◻ fly as pilot-in-command of a gyroplane at night unless a Night Rating (Gyroplanes) is held and he has, within the preceding 13 months, carried out as pilot-in-command not less than five take-offs and five landings at night. Night is defined for these purposes as at a time when the depression of the centre of the sun is not less than 12° below the horizon.

The holder of either a UK PPL (Helicopters) or JAR PPL (Helicopters) must not:

◻ fly as pilot-in-command of a helicopter at night unless his licence includes a Night Rating (Helicopters) or a Night Qualification (Helicopters) and he either:
  – holds an instrument rating, or
  – within the preceding 90 days has made three circuits, each to include take-offs and landings by night as the sole manipulator of the controls of a helicopter of the same type.

NOTE While the requirement for currency by day is that the three circuits must be solo, at night the requirement is met by dual circuits.

Validity of a PPL(H) is maintained by passing a proficiency check each year on each type included on the licence and also by flying a total of 2 hours on each type. The proficiency check may be included in calculating the 2 hours. There are some exceptions to single-engine piston helicopter types which fall into a common group listed in JAR FCL2.

In order to qualify for issue of the PPL(H), a cross-country flight of 100 nm must be completed, including two landings at licensed landing sites.

Proficiency check is a demonstration of continued knowledge and skill for the revalidation of a licence or rating. The proficiency check includes such oral examination as the examiner may determine.

To obtain a Night Qualification the holder of a PPL(H) shall have completed at least 100 hours of flight time as pilot of helicopters after the issue of the licence, including at least 60 hours as pilot-in-command of helicopters and 20 hours cross-country flight.

PPL (Balloons and Airships)

There are some notable differences in the training required to obtain a PPL (Balloons and Airships).

Contact the British Balloon and Airships Club (BBAC) at their website: www.bbac.org

▢ No medical examination is required. It is sufficient for the applicant to make a declaration of fitness on form FCL150/AB (which can be obtained from the CAA medical branch). This needs to be countersigned by the applicant's GP.

▢ The minimum flying experience for the grant of the licence is 16 hours under instruction, comprising at least six flights. Four of these flights must be made with an approved instructor of the British Balloon and Airships Club (BBAC), but the remainder may be carried out with any licensed balloon pilot. It is recommended that this pilot has a minimum of 10 hours as pilot-in-command of balloons. The student must keep a logbook and a training record in which each flight is recorded and signed by the instructing pilot.

▢ The written examinations comprise: air law, navigation, airmanship and balloon systems, meteorology, and human performance and limitations. The pass mark for these papers is 70%.

▢ In addition to a flight test with an examiner, the applicant must have completed a solo flight of at least 30 minutes under the supervision of the examiner or a delegated instructor, and also have carried out at least one tethered flight supervised by any qualified balloon pilot.

▢ All of the requirements for issue of the licence must be completed within the 12 month period immediately preceding the application for the licence.

A PPL holder may fly as pilot-in-command of any aircraft type specified in Part 1 of the aircraft rating in the licence and co-pilot of any type specified in the rating provided he has made a valid declaration of fitness and does not fly for the purpose of public transport or aerial work, other than aerial work which is instruction in flying or the conducting of flying tests.

▢ In either case the aerial work would have to be done in an aircraft owned, or operated under arrangements entered into, by a flying club of which the person instructing or conducting the test and the person being instructed or undergoing the test are both members.

- He shall not receive any remuneration for his services as a pilot on a flight other than remuneration for the giving of such instruction or flight tests as specified above.

- Although PPL (Balloons and Airships) licence-holders are not required by law to hold a valid Certificate of Test or Certificate of Experience, they are not entitled to exercise the privileges of the licence unless during the previous 13 months they have made 5 flights, each of not less than 5 minutes' duration, as PIC in a free balloon, or have satisfactorily completed a flight test.

The National Private Pilot's Licence (NPPL)

The NPPL has been introduced with effect from 30 July 2002. The CAA has supported its introduction under the auspices of a steering committe comprising representatives of various industry bodies. The NPPL is administered by the National Pilots Licensing Group Limited with delegated authority from the CAA.

Minimum Requirements

The NPPL is a simpler form of licence than the JAR PPL and the minimum requirements for its issue are as follows:

- 22 hours of dual training, including 1 hour of instrument appreciation.
- 10 hours of solo flight, including one cross-country flight in which two full-stop landings at two aerodromes other than the aerodrome of departure shall be made.
- The theoretical knowledge requirement for the NPPL will initially follow the JAR-FCL requirements listed on page 129 of this manual. In due course it is envisaged that separate written examinations of a simpler format will be introduced.
- A Navigation Flight Test (NFT) and a General Flight Test (GFT), each of a minimum of one hour duration, shall be passed; the NFT prior to the solo qualifying cross-country flight.
- Minimum age for first solo is 16, for licence issue is 17.
- The minimum medical standards to fly solo are to be the equivalent of the Driver and Vehicle Licensing Agency (DVLA) Group 1 car driving medical standards. This can be certified by your own general practitioner (GP).
- To carry passengers after the issue of the licence you will require a standard of medical fitness equivalent to DVLA Group 2 professional driving medical standard. Again your GP can verify this.

Unless differences training has been completed and signed off in the pilot's logbook the holder of an NPPL may not:

- pilot an aeroplane with more than 4 persons on board;
- pilot an aircraft fitted with features such as retractable undercarriages or variable-pitch propellers.

Note also the following limitations:
- ☐ The licence is limited to day VFR conditions with a minimum in-flight visibility of 5 km.
- ☐ The privileges of the NPPL are restricted to UK airspace.
- ☐ An NPPL holder may accept a Special VFR clearance for flight in a control zone in flight visibility of 10 km or more.
- ☐ An NPPL holder may not fly an aircraft with a maxium continuous cruising speed in excess of 140 kt IAS.

Upgrading to JAR PPL

An NPPL may be upgraded to a JAR-FCL PPL with a credit of up to 30 hours provided all flying training has been completed by a JAR-qualified instructor and all JAR PPL ground examinations have been passed.

A further 15 hours of training will need to be completed, of which 5 hours may be undertaken in a simulator.

As these changes are still in a state of flux, you should check for new developments with your flying school or the National PPL website (www.nppl.uk.com).

Small Light Aeroplanes and Microlights

Microlights are classified as follows:

Contact the British Microlight Aircraft Association (BMAA) at tel: 01869 33888

- ☐ two-seat
- ☐ single-seat
- ☐ three-axis
- ☐ weight-shift

Microlight aircraft are designed to carry not more than two people and have a maximum authorised total weight not exceeding:

- ☐ 300 kg for a single-seat landplane (or 390 kg for a single-seat landplane for which an individual UK Permit to Fly or Certificate of Airworthiness was in force prior to 1 January 2003); or
- ☐ 450 kg for a two-seat landplane;
- ☐ 495 kg for a two-seat amphibian or floatplane;
- ☐ 330 kg for a single-seat amphibian or floatplane.

Microlights are limited to a wing loading not exceeding 25 kg per square metre and a stalling speed (V_{S0}) of 35 kt calibrated airspeed.

A minimum of 25 hours of dual instruction and 10 hours of solo flight is required before each licence issue. However, a restricted licence is available after 15 hours of training of which 7 hours must be solo. The restricted licence limits the holder to flight within an 8 mile radius of the departure airfield.

Microlight aircraft are operated on a Permit to Fly and no aerial work is permitted other than instruction.

An existing UK or JAR-FCL PPL(A) holder who has national microlight privileges together with other aircraft ratings may choose to maintain only the microlight rating, provided that the licence includes a valid JAR-FCL Class 2 medical certificate. A declaration of health would not be sufficient to keep the licence valid unless the microlight rating was the only rating on the licence.

If an applicant wishes to fly only microlights, a PPL(A) (Microlight) can be issued provided it is the only rating on the licence. It is not possible to add any other rating to a PPL(A) (Microlight). Therefore, the holder of a PPL(A) (Microlight) who wishes to fly single-engined piston (SEP) aeroplanes must obtain a JAR-FCL PPL(A) with SEP class rating. It is not possible to 'downgrade' an existing PPL(A) to fly only microlights, i.e. maintaining only a declaration of health. An application has to be made for the issue of a separate national UK microlight licence.

After undergoing a Microlight General Flight Test (GFT) with a microlight examiner, the validity of the rating may be maintained by Certificate of Experience, by flying a minimum of 5 hours in a microlight every 13 months.

Ratings and Maintaining a Rating

The Private Pilot's Licence will include one or more **class ratings** specifying the aircraft classes that the holder may fly and any additional privileges that the holder has obtained.

Aircraft ratings entitle holders to pilot aircraft of the type or class specified (provided that a valid and appropriate Certificate of Revalidation is held). For PPL holders, aeroplanes are subdivided into the following aircraft class rating groups:

- single-engine piston landplanes
- single-engine piston seaplanes
- multi-engine piston landplanes
- multi-engine piston seaplanes
- touring motor gliders (TMGs)

Single-Engine Piston Landplane

Normally the first *class rating* to be obtained, JAR-FCL PPL(A) includes virtually all single-engine piston aircraft irrespective of weight.

The JAR-FCL PPL(A) requires a minimum of 45 hours training, of which a minimum of 25 hours dual instruction and 10 hours solo flight is required. Solo flight time must include 5 hours of cross-country flying and one flight must be of at least 150 nautical miles, with two intermediate full-stop landings made away from the departure field. Solo flight may be undertaken from the age of 16, but the PPL will not be issued before the applicant reaches 17 years of age.

Holders of pilot licences or equivalent privileges for helicopters, microlights (three-axis), gliders or SLMG may be credited with 10% of their total pilot-in-command flying time of such aircraft up to a maximum of 10 hours, towards a JAR PPL(A).

Training must be conducted at an Approved Flying Training Organisation (FTOs) or at a Registered Facility (RF).

NOTE Pilots who later wish to fly more complicated aircraft equipped with variable-pitch propellers, retractable undercarriages, turbo or supercharged engines, or aeroplanes with pressurised cabins, are required to undergo differences training with a flight instructor, and such training must be recorded in the pilot's logbook. Tailwheel aircraft also require differences training if you have learnt on a nosewheel/tricycle undercarriage aircraft.

Single-Engine Piston Seaplane

This is a separate class rating with training requirements very similar to a SEP (Land) rating, with an additional seamanship written examination.

> *Skill Test* is a demonstration of knowledge and skill of initial licence issue, rating issue or renewal. The test includes such oral examination as the examiner may determine.

This class rating is normally added to an existing single-engine class rating (land) at approved JAR facilities by JAR-qualified instructors, along with a written seamanship examination and skill test. Flying training consists of a minimum of 5 hours dual training from a sea or water facility.

Alternatively, training may be completed in another ICAO country, such as the USA, and added to an FAA PPL, and then added to an existing JAR PPL. Check that the training syllabus is compatible with JAR-FCL requirements if you take this course.

Multi-Engine Piston Landplanes

This class rating is normally added to an existing single-engine class rating and includes virtually all single-pilot multi-engine piston aircraft of this class, irrespective of the aircraft's weight. It replaces the old 'B' Rating, which was limited to a maximum all-up weight of 5,700 kg.

The applicant must have at least 70 hours 'pilot-in-command' experience (aeroplanes), before the issue of the rating. The multi-engine course comprises a minimum of 2½ hours of two-engine flight, 3½ hours asymmetric flight training, 7 hours of theoretical knowledge of multi-engine operation and a skill test with a flight examiner. In addition, a multi-choice written examination relating to the specific aircraft type and general multi-engine operations must be passed. Note that this course may only be conducted at a Flight Training Organisation (FTO).

Multi-Engine Piston Seaplane

This is yet another separate class rating requiring theoretical and dual training, a written examination in multi-engine operation and seamanship (if not already completed), plus a class rating skill test.

Self-Launching Motor Gliders (SLMGs) and Touring Motor Gliders (TMGs)

Training for TMGs and SLMGs is very similar to that for single-engine piston aeroplanes (with some exceptions such as starting and stopping the engine in flight) and differentiates between TMGs and SLMGs (which are considered to be powered sailplanes (gliders) that can retract the engine in flight).

Aeroplanes categorised as TMGs such as Fournier, Grob, Scheibe and others are listed in the JAR-FCL1 manual (AMC FCL 1-215).

The National SLMG rating confers the privilege to fly any motor glider, including those that meet the JAR definition of touring motor gliders, in UK airspace.

The National rating is valid for 13 months and is revalidated with an examiner's stamp in your logbook (subject to the meeting of experience requirements of 5 hours in the preceding 13 months or passing a skill test).

JAR-FCL TMG CLASS RATING In order to convert a National PPL(A) SLMG rating to a JAR PPL(A) TMG class rating, the applicant requires:

☐ a minimum of 75 hours as pilot of aircraft meeting the JAR-FCL definition of TMG (see page 127);

☐ have a current CofT or Certificate of Revalidation for such aircraft;

☐ hold a JAR-FCL Class 2 medical certificate;

☐ demonstrate knowledge of the relevant parts of JAA requirements (this may be satisfied by signing a declaration contained in the application form).

NOTE If the holder of a National PPL(A) SLMG wishes to fly single-engine (SEP) he will have to complete a full JAR PPL(A) course and will only obtain a maximum of 10 hours' credit towards the 45 hours required. It is expected that the UK PPL SLMG course will be replaced by the NPPL SLMG in due course.

The JAR-FCL Touring Motor Glider class rating is issued and maintained in accordance with JAR-FCL requirements and is very similar to those required for the maintenance of the SEP class rating.

Training for the JAR-FCL PPL, whether for single-engine piston (SEP) or TMG may only be undertaken by a CAA-registered facility or approved flying training organisation, at a CAA-licensed airfield. The aircraft used for training must have a Public Transport Certificate of Airworthiness.

Contact the British Gliding Association (BGA) at tel: 011625 31051

At the time of writing, the British Gliding Association (BGA) has been granted an exemption by the CAA to continue training for the UK PPL SLMG at a restricted number of gliding clubs, provided the trainee holds a British Gliding Association Glider Pilot's Licence prior to commencing training. This training may take place in 'non-TMG SLMGs', and does not require a licensed airfield or a Public Transport category CofA. This arrangement is to allow for the training of pilots to become BGA SLMG instructors, and for those who wish to fly self-launching sailplanes. Further information is available from the BGA.

Certificate of Test and Certificate of Revalidation for JAR-FCL Single-Engined Piston Aircraft and Touring Motor Gliders (TMG)

A Certificate of Test (CofT) is valid for 24 months from the date of the Flight Test which it certifies and may be revalidated by passing a Proficiency Check with a flight examiner within a period of 3 months preceding the date of expiry or by completing a Certificate of Revalidation.

A Certificate of Revalidation (CofR) relates to experience gained within the last 12 months preceding its expiry. The applicant must record at least 12 hours logged flight time, including a minimum of 6 hours as pilot-in-command (P1), 12 take-offs and landings and a 1-hour training flight with a flight instructor which must be recorded and signed in the applicant's logbook.

In all cases it is the holder's responsibility to ensure that revalidations are made within the appropriate time or experience limit, if licence privileges are to be maintained. Revalidations (by experience) may be completed at any time in the 12 months preceding the expiration date of the rating.

If the 24 months (but not more than 5 years) are exceeded without a Certificate of Revalidation extending the licence, a Proficiency Check with an examiner is required to renew the privileges.

Multi-Engine Piston Class Rating for Single Pilot Aircraft (the old National 'B' rating)

To maintain this rating applicants will need to complete:

☐ a minimum of ten route sectors (flights), including arrival, approach and landing phases, during the preceding 12 months (a route sector is defined as a flight of at least 15 minutes in the cruise plus departure and arrival); **and**

☐ a Proficiency Check with a flight examiner within the 3 months preceding the expiry date of the rating.

Should the rating expire before meeting CofR requirements, then a Proficiency Check **and** one route sector with an examiner is required to renew the privileges of the rating.

Type Ratings

A type rating requirement is established for aeroplanes of significant complexity of handling characteristics and is assessed on the following criteria:

☐ the aeroplane has handling characteristics that require additional flying (or simulator) training; or

☐ the aeroplane requires a separate Airworthiness Type Certificate.

Type ratings are established for all types of single-pilot, single-engine or multi-engine turbo propeller (or jet) aeroplanes or for any aeroplane requiring two or more pilots. The CAA may establish a type rating for any other type of aeroplane if it is considered to be necessary.

A list of specific type ratings is contained in JAR-FCL (AMC FCL 1.220).

NOTE Type ratings are valid for 12 months.

IMC Rating (Aeroplanes)

The IMC Rating entitles holders to act as pilot-in-command of an aeroplane without being subject to the flight visibility restrictions stated earlier, provided that:

1. They do not fly on a Special VFR (SVFR) flight in a Control Zone in a flight visibility of less than 3 km.

2. They do not act as pilot-in-command when the aeroplane is taking off or landing at any place if the flight visibility below cloud is less than 1,800 m.

3. It is only used in UK territorial airspace unless written permission is obtained from another State. (The IMC Rating is a national rating.)

4. It is not used in controlled airspace requiring compliance with Instrument Flight Rules (IFR).

NOTE An IMC holder may accept an IFR clearance in IMC conditions into or through Class D or E airspace.

Instrument Rating (Aeroplanes)

The Instrument Rating entitles the holder to be pilot-in-command or co-pilot of an aeroplane flying in Class A, B or C airspace in circumstances which require compliance with the Instrument Flight Rules (IFR).

Night Rating (Aeroplanes)/Night Qualification

The Night Rating/Qualification entitles the holder to be pilot-in-command at night of an aeroplane in which a passenger is carried. Flights at night require observance of Rule 29 (IFR: Minimum Height) and Rule 30 (IFR: Quadrantal Rule and Semicircular Rule).

Flight Instructor's Rating

The Flight Instructor's Rating entitles the holder to give instruction in flying (in the type or class specified in the rating) to any person who is flying with him for the purpose of becoming qualified for the grant of a pilot's licence or the inclusion or variation of any rating in the licence.

Restricted Flight Instructor's Rating

The Restricted Flight Instructor's Rating entitles the holder to instruct in flying aircraft of types or classes specified in the rating, provided that:

1. Instruction is given under the supervision of a holder of a Flight Instructor's Rating present during the take-off and landing at the aerodrome at which the instruction is to begin and end; and

2. No direction or authorisation is given to students regarding their first solo flight or first solo cross-country flight, whether by day or by night (a cross-country flight being one that is properly pre-planned with a route, turning points and destination, proceeding more than 3 nm from the departure aerodrome, and of suitable duration and distance as to require in-flight tracking adjustments).

INSTRUCTION IN FLYING Flight instruction for the grant of a licence or rating must only be given by those who hold a licence entitling them to act as pilot-in-command for that purpose and in the circumstances under which instruction is to be given.

Instrument Rating (Helicopters)

This rating entitles the holder to be pilot-in-command or co-pilot of a helicopter in Class A, B or C airspace in circumstances which require compliance with Instrument Flight Rules (IFR).

Night Rating (Helicopters & Gyroplanes)

This rating entitles the holder to be pilot-in-command at night of a helicopter or gyroplane in which a passenger is carried.

Personal Flying Logbooks

A personal flying logbook in a suitable format (this may include computerised versions in certain circumstances) must be kept by qualified pilots and those flying to obtain or renew a licence. The following particulars shall be recorded:

- ☐ the name and address of the licence holder;
- ☐ the details of the licence;
- ☐ the details of each flight made as a member of a flight crew or for the purpose of obtaining or renewing a licence, including:
  - the date and places at which the holder embarked and disembarked and the time spent during the flight in either of the capacities mentioned (flight time being from when the aircraft first moves under its own power until the moment it comes to rest after landing);
  - the type and registration marks of the aircraft;
  - the capacity in which the holder acted in flight (with a flight instructor on board, the time is logged as *dual;* when solo or pilot-in-command, the time is logged as *in command*);
  - details of any special conditions including night flying, instrument flying and in particular the training exercises specified in the PPL qualifying requirements;
  - particulars of any test or examination undertaken whilst in flight;
  - particulars of flight simulator tests.
- ☐ A helicopter is deemed to be in flight from the moment it first moves under its own power for the purpose of taking off until the rotors are next stopped.
- ☐ Particulars of any test or examination undertaken whilst in a flight simulator shall be recorded in the logbook, including:
  - the date;
  - the type of simulator;
  - the capacity in which the holder acted; and
  - the nature of the test or examination.

It is worth noting that there is an offence in the ANO of intentionally damaging, altering or mutilating entries in a logbook or licence. There is evidence of the CAA successfully prosecuting pilots for such offences (which amount to forgery). The penalty may be a fine or even imprisonment.

Your logbook is a valuable legal document and is your only proof of the true record of your flying so you should take care of it. It is worth photocopying the pages from time to time in case you do lose it, since in such a case the CAA will require you to make a witness statement certifying its loss and to attempt to reconstruct the hours claimed.

NOTE The *commander* of an aircraft is required to produce his logbook within a reasonable time after being requested to do so by an authorised person, for up to **2 years** from the date of the last entry.

Date	AIRCRAFT		CAPTAIN	Holder's Operating Capacity	JOURNEY or Nature of Flight		Depart-ure (G.M.T.)	Arrival (G.M.T.)
	Type	Registra-tion			From	To		
AUG 2	AA-5A CHEETAH	G-KILT	SELF	P1	ELSTREE	ELSTREE	12.20	13.50
AUG 8	SLINGSBY T67	G-BKTZ	SELF	P1	LEAVESDEN	BIRMINGHAM	07.30	08.55
AUG 8	"	G-BKTZ	SELF	P1	BIRMINGHAM	CRANFIELD	12.45	13.15
AUG 8	"	G-BKTZ	SELF	P1	CRANFIELD	LEAVESDEN	17.00	17.45
AUG 24	AZTEC	G-ARBN	P. GODWIN	P1 US	ELSTREE	LOCAL	12.05	13.15

75 25	32 45				02 15			05 30	Hrs. Mins.	Totals Brought Forward
(1)	(2)	(3)	(4)	(5)		(6) FLYING TIMES (7)	(8)	(9)	(10)	

DAY				NIGHT				Instrument Flying	Simulated Instrument Flying	REMARKS
Single-Engine		Multi-Engine		Single-Engine		Multi-Engine				Including counter-signature for P.1/S
In Command	Dual or P.2	In Command	Dual or P.2	In Command	Dual or P.2	In Command	Dual or P.2			
1.30										
1.25										
.30										
.45										
		1.10								

■ *Figure 10-3* **Typical logbook entries**

Now complete **Exercises 10 – Pilots' Licences.**

Operation of Aircraft

Ops Procedures (UK ANO)

The ICAO standards and recommended practices for the international operation of general aviation aircraft contained in Annex 6 differ in some parts from the ANO's requirements. Therefore we set out those parts of the ANO regarding operation of aircraft first, followed by the relevant procedures relating to international flights which might be made by PPL holders (page 155).

Pre-Flight Actions

Pre-Flight Actions of the Pilot-in-Command (ANO Article 43)

The pilot-in-command must be satisfied before each flight:

- [ ] that the flight can be made safely, taking into account the latest information available as to route and aerodromes to be used, weather reports and forecasts available, and options open if the flight cannot be completed as planned;

- [ ] that equipment (including radio) required by regulation is carried and working and that appropriate maps and charts and navigational equipment is carried;

- [ ] that the aircraft is fit for the flight and has a valid Certificate of Maintenance Review where required;

- [ ] that any load is safe in terms of weight, distribution and security;

- [ ] that enough fuel, oil, coolant and ballast (if appropriate) is carried, including a margin for safety;

- [ ] that, with regard to its performance, the aircraft is capable of safely taking off, reaching and maintaining a safe height and making a safe landing at the intended destination. (In very approximate terms, a 10% increase in weight will increase the **take-off distance required** by 20%; a tailwind component of 10% of the take-off speed – usually about 5 knots – will increase take-off distance required by 20%; a 10°C rise in temperature or 1,000 ft rise in elevation will increase take-off distance required by about 10%. Precise consideration, of course, requires reference to the take-off and landing charts or tables for your particular aeroplane);

- [ ] that all required pre-flight checks have been carried out.

Passenger Briefing (ANO Article 44)

Before taking off on any flight, the pilot-in-command must brief the passengers on the position and method of use of:
- emergency exits;
- safety belts or safety harnesses; and
- oxygen equipment and life jackets (when they are required to be carried).

Air Traffic

Authority of the Commander of Aircraft (ANO Article 67)

Everyone on board must obey all lawful commands which the aircraft commander gives for the purpose of the safety of the aircraft and of people or property carried, or for the safety, efficiency or regularity of air navigation.

Crew Composition (ANO Article 20)

An aircraft must not fly unless the crew composition meets the legal requirements of the country of registration. (UK requirements are stated in ANO Article 20.)

Pilots to Remain at Controls (ANO Article 41)

The commander of a flying machine or glider is responsible for ensuring that:
- one pilot is at the controls at all times in flight;
- both pilots are at the controls during take-off and landing if the aircraft is required to carry two pilots under the Order;
- each pilot at the controls wears a safety belt or a safety harness.

NOTE A harness must be worn during take-off and landing when one is required to be provided by Article 14.

Method of Carriage of Persons (ANO Article 61)

No one must be in or on any part of an aircraft in flight not designed for the accommodation of people, nor be in or on any object (other than a glider or flying machine) towed or attached to an aircraft in flight, except for temporary access to any part of an aircraft:
- for the purpose of taking action necessary for the safety of the aircraft or of any person, animal or goods on board;
- in which cargo or stores are carried and which is designed to enable a person to have access during flight.

Smoking in Aircraft (ANO Article 66)

Notices indicating when smoking is prohibited in any part of an aircraft must be exhibited so as to be visible from each passenger seat and obeyed.

It is good airmanship to observe *no smoking* during take-off, landing and low flying, although smoking is not specifically prohibited at any time except when the 'no smoking' sign is displayed. The pilot has the authority to ban smoking at all times, bearing in mind that cigarette smoke in the cockpit may impair performance and cause distress to certain people.

Drunkenness in Aircraft (ANO Article 65)

A person shall not enter an aircraft when drunk, or be drunk on any aircraft. A person must not, when acting as a crew member or being carried for the purpose of so acting, be under the influence of drink or a drug to such an extent as to impair his capacity so to act. Although no exact times are specified in the ANO, it is reasonable for a pilot not to fly until at least 8 hours after any alcohol has been imbibed and, if excessive amounts have been consumed, for this time to be considerably extended.

Acting in a Disruptive Manner (ANO Article 68)

Following publicity in the media about 'air rage', a new offence was introduced in ANO 2000: "A person shall not use threatening or abusive or insulting language towards a member of the crew, nor shall they behave in a threatening, abusive, insulting or disorderly manner towards a crew member, nor interfere intentionally with the performance of the duties of the crew."

Imperilling the Safety of Aircraft, Persons or Property (ANO Articles 63 and 64)

A person must not, recklessly or negligently, either act in a manner likely to endanger an aircraft or anyone in it, or cause or permit an aircraft to endanger anyone or any property; e.g. a person deliberately damaging an aeroplane or a pilot flying excessively low, carrying inadequate fuel, etc., would be in contravention of this Article.

Operation of Radio in Aircraft (ANO Article 46)

Anyone operating an aircraft's radio equipment must hold an appropriate radiotelephony (R/T) licence (see Chapter 10). The radio equipment itself must be licensed and operated only in accordance with the conditions of that licence.

Flight Recording Systems (ANO Article 53 and Schedule 4)

Although this article will not usually apply to a private pilot, aeroplanes over 5,700 kg and helicopters over 2,700 kg maximum take-off weight authorised are required, in certain circumstances, to carry and operate flight recording systems, and to preserve the records from them for a specified time.

Towing Gliders (ANO Article 54)

An aircraft in flight may not tow a glider unless its Certificate of Airworthiness includes an express provision that it may. The Article also specifies safety requirements with regard to signals, take-off techniques, and length and condition of tow-ropes.

Towing, Picking Up and Raising of Persons and Articles (ANO Article 55)

An aircraft in flight may not tow, pick up or raise anything unless expressly permitted by its Certificate of Airworthiness. The Article also contains specific requirements concerning the several types of activity which fall under its heading, e.g. banner-towing and loads slung from helicopters.

Dropping of Articles and Animals (ANO Article 56)

Articles and animals whether or not attached to a parachute must not be dropped, projected or lowered so as to endanger people or property and should not be dropped, projected or lowered in other circumstances without **written CAA approval**, except for certain articles in certain situations, e.g. dropping a life-raft to save lives, or jettisoning of items to lighten the load in an emergency.

Pilots involved in crop-spraying operations are directed to Article 58 and the requirement to obtain an *Aerial Application Certificate*. Also, if the Certificate of Airworthiness for a helicopter expressly allows it, then the lowering of any person, animal or article to the surface is permitted.

Dropping of Persons (ANO Article 57) – (Parachuting)

Except for escaping in an emergency, no one must drop to the surface or jump from an aircraft in flight without the **written permission of the CAA** and then only in such a manner as not to endanger people or property. In the case of parachuting, the Certificate of Airworthiness of the aircraft must include a provision that it may be so used and the aircraft must be operated in accordance with the Parachuting Manual which the holder of a *CAA Permission to Drop* has to maintain.

Carriage of Weapons and Munitions of War (ANO Article 59)

No weapon may be carried on board an aircraft unless:
- the consent of the operator is obtained;
- the weapon is unloaded in the case of a firearm; and
- the weapon is carried in a part of the aircraft not accessible to passengers.

Munitions of war (including weapons, ammunition, explosives or noxious substances) shall not be carried without the written permission of the CAA and then only in accordance with any conditions relating thereto.

Carriage of Dangerous Goods (ANO Article 60 and Annex 18)

The Secretary of State can classify certain articles and substances as **dangerous goods** (see page 178 for definition). As a pilot, you should be aware that goods which may be fairly innocuous at ground level and/or in the open air, may be very dangerous in flight where atmospheric pressure is reduced and persons are restricted to the confined space of the aircraft cabin.

Many dangerous goods are not permitted in aircraft and those permitted to be carried by air are subject to special conditions of handling, loading, packing, labelling and documentation. Classification lists and conditions are published in the form of regulations made from time to time by the Secretary of State.

As a commonsense practice, private pilots should make a habit of checking the items carried by passengers in their aircraft for obviously dangerous items, such as certain types of matches and lighters, any compressed gases, corrosives or flammable substances. Examples include aerosol hairsprays, batteries, camping gas cylinders and fire lighters, all of which might be carried on that weekend camping trip! As a general rule, *if in doubt, leave it out!*

In addition, the use of a mobile telephone on board an aircraft can interfere seriously with its navigational equipment. Even in stand-by mode, mobile phones emit signals periodically. Not only is there an adverse effect on safety, but the use of mobile telephones in an aircraft is a breach of the telephone user's licence. Therefore the commander of an aircraft should ensure that all mobile telephones are switched off prior to engine start. See AIC 62/1999 (Pink 196).

Rules of the Air (ANO Article 84, previously 74)

This Article is the legal basis for the large subject *Rules of the Air and Air Traffic Control Regulations.* These are the 'rules' referred to throughout this text, mainly in Chapter 2. Article 84 also lists when the Rules of the Air may be departed from, and to what extent.

Balloons, Kites, Airships, Gliders and Parascending Parachutes (ANO Article 86)

There are specific limitations for the flying of captive balloons and kites with regard to the method of mooring, location and height. A captive balloon or kite shall not be flown:

- ▢ at a height of more than 60 metres agl (above ground level);
- ▢ within 60 metres of any vessel, vehicle or structure;
- ▢ within 5 km of an aerodrome.

A balloon, either captive or free, exceeding 2 metres in any linear dimension (including any basket, etc.) must not be flown in controlled airspace notified for the purposes of this restriction.

An airship shall not be moored.

A glider or parascending parachute shall not be launched by winch and cable, or by ground tow, to a height of more than 60 metres above ground level.

NOTE Each of these limitations may be varied by request and with written permission of the CAA, subject to conditions which will accompany the permission.

Documentation and Records

Documents to be Carried (ANO Article 76 and Schedule 11)

On an **aerial work** flight, the documents to be carried are:

- the aircraft's radio licence;
- the Certificate of Airworthiness (CofA);
- flightcrew licence(s);
- one copy of each Certificate of Maintenance Review in force;
- the Technical Log (if any and if required);
- the Certificate of Registration (if the flight is international, i.e. beyond the bounds of the UK, Channel Islands and Isle of Man);
- a copy of the notified procedures to be followed if the aircraft is intercepted, and the visual signals for use in such circumstances (this applies to international flights only); and
- although not always a requirement, it is good practice to carry the insurance documents.

NOTE 1 The requirement to carry full documentation on an aerial work flight is **waived** when it is intended to take off and land at the **same aerodrome** and remain **within UK airspace**. Also, with the CAA's approval a Flight Manual need not be carried if an Operations Manual, which contains the specified information, is carried.

NOTE 2 A **private** flight need only carry the first five documents listed above if international air navigation is involved, but for flights totally within the confines of the UK, Channel Islands and the Isle of Man no documents need be carried.

NOTE 3 The requirement to carry insurance documents is mandatory in Spain, and a translation of the insurance certificate must be available in Spanish.

Production of Documents (ANO Article 78)

The commander of an aircraft shall, within a reasonable time after being requested to do so by an authorised person (a constable, or anyone authorised by the Secretary of State or the CAA in a manner appropriate to the case), produce:

☐ the Certificates of Registration and Airworthiness in respect of the aircraft;

☐ the licence(s) of its flight crew and the personal flying logbooks (for up to two years after the date of the last entry in any book); and

☐ such other documents that the aircraft should carry when in flight.

Offences in Relation to Documents & Records (ANO Article 83)

It is an offence to:

☐ use a forged, altered, revoked or suspended document issued under the Order, or to use one to which you are not entitled;

☐ lend a document issued under the Order, or otherwise allow it to be used by anyone else, with intent to deceive;

☐ forge or render illegible any logbook or record required by the Order, or to make a false entry or material omission, or to destroy it during the period for which it is required to be preserved (all entries in such documents must be made in ink or indelible pencil).

NOTE The CAA can revoke or vary any licence or document that it has issued.

Operation of SSR Transponders

The airborne component of the ATC secondary surveillance radar (SSR) system is known as the **transponder**, and there are requirements prescribed in the ANO and AIP ENR 1-6-0 for carriage of SSR transponder equipment.

The two capabilities of transponders are:

☐ Mode A – capable of selecting 4,096 4-digit codes; and

☐ Mode C – 4,096 codes, with the capability of transmitting the aircraft's altitude (read from a suitable linked altimeter known as an *encoding altimeter*).

Mandatory Requirements

☐ A transponder operating on Mode A (at least) is required in the Scottish TMA between 6,000 ft amsl and FL100.

☐ Mode A and Mode C with altitude reporting is required for:

– the whole of UK airspace at and above FL100; and

– in UK controlled airspace notified for the purposes of ANO Schedule 5 (sub-paragraph 2(1)(a)) below FL100 when operating under Instrument Flight Rules (IFR), except when receiving a 'crossing service'. (The ANO Schedule 5 notified airspace is listed in AIP ENR 1-4.)

Gliders are exempt from the above requirements.

In airspace where the use of transponders is *not* mandatory, pilots should *squawk* (the term for using a transponder code) the Conspicuity Code (see below) except when remaining within the confines of an aerodrome traffic pattern below 3,000 ft agl.

Transponder Codes

The conspicuity code 7000 (and Mode C if so equipped) shall be used for normal operations *at and above FL100* except:

- when receiving a service from an ATSU or ADU requiring a different transponder code setting; or
- when circumstances require the use of one of the special purpose codes (see below).

Below FL100, pilots should select the conspicuity code and Mode C, except as above. Special-purpose codes are reserved for various purposes as described in the table below.

SPECIAL-PURPOSE TRANSPONDER CODES	
Code 7700	*to indicate an emergency condition (except that, if the aircraft is already transmitting a code and receiving an air traffic service, that code will normally be retained)*
Code 7600	*to indicate a radio failure*
Code 7500	*to indicate unlawful interference*
Code 7000	*conspicuity code for normal operations*
Code 2000	*when entering UK airspace from an adjacent region where transponder operation has not been required*
Code 7004	*a special code to indicate that training, practising or displaying of aerobatics is taking place*
Code 0033	*a special code to be selected 5 minutes before parachute dropping begins and until the parachutists are estimated to have reached the ground*

Mode C (if so equipped) should be operated with all of the above codes.

NOTE When selecting the conspicuity code 7000, pilots should be careful not to activate inadvertently a special-purpose code (7700, 7600 or 7500).

For further reading on transponder operations, see AIP ENR 1-6-2, and Vol. 7 of *The Air Pilot's Manual*.

Airprox Reporting Procedure

When a pilot considers that the safety of his or her aircraft may have been endangered by the proximity of another aircraft, within UK airspace, the incident must be reported as an *Airprox*. This is a requirement stated in Article 117 (*Mandatory Reporting*) of the ANO.

An initial report should be made by radio to the Air Traffic Service Unit (ATSU) with which the aircraft is in communication. If it is impossible to report by radio, a report should be made by telephone or other means to any ATSU, but preferably to an Air Traffic Control Centre (ATCC) immediately after landing. Reports made by telephone or radio must be confirmed within seven days on CA Form 1094.

AIC 15/1999 (Pink 186) *Airprox Reporting − UK and Foreign Airspace* and AIP ENR 1-14 contain information on Airprox Reporting.

International Flights (ICAO Annex 6)

The following extracts from Annex 6 cover matters particularly relevant to PPL holders making international flights.

General

3.1 The pilot-in-command shall comply with the relevant laws, regulations and procedures of the States in which the aircraft is operated.

3.2 The pilot-in-command shall be responsible for the operation and safety of the aeroplane and for the safety of all persons on board, during the flight.

3.3 Should an emergency situation occur which endangers the safety of the aeroplane or people, and requires the pilot to take action which violates local regulations or procedures, the pilot shall notify the appropriate authority as soon as possible. Some States may require the pilot to submit a report on the violation, normally within ten days.

3.4 In the event of an accident involving the aeroplane which results in serious injury or death or substantial damage to the aeroplane or property the pilot-in-command shall be responsible for notifying the appropriate authority as quickly as possible.

3.5 ICAO recommends that the pilot-in-command should carry on board the aeroplane essential information on search and rescue services in the areas over which the aeroplane will be flown.

Adequacy of Operating Facilities

4.1 The pilot in command shall not begin a flight unless he has ascertained that the aerodrome facilities, communication facilities and navigation aids required are adequate for the safe operation of the aeroplane.

Aerodrome Operating Minima

4.2 The pilot-in-command shall not fly below the operating minima specified for an aerodrome, except with State approval.

Briefing

4.3.1 The pilot-in-command shall ensure that crew members and passengers are briefed on the location and use of:
- ☐ seat belts;
- ☐ emergency exits;
- ☐ life jackets;
- ☐ oxygen equipment;
- ☐ any other emergency equipment, including passenger briefing cards.

4.3.2 The pilot-in-command shall ensure that everyone on board is familiar with the location and use of emergency equipment carried for collective use, such as life rafts.

Aeroplane Airworthiness and Safety Precautions

4.4.1 The pilot-in-command shall not begin a flight unless he is satisfied that:
- ☐ the aeroplane is airworthy, registered and has the appropriate certificates on board;
- ☐ the instruments and equipment in the aircraft are appropriate to the expected flight conditions;
- ☐ necessary maintenance has been completed;
- ☐ the aeroplane's weight and balance will be within safe limits for the flight;
- ☐ cargo is correctly stowed and secured;
- ☐ the aeroplane's operating limitations, as described in the Flight Manual, will not be exceeded.

NOTE ICAO recommends that the pilot-in-command should have sufficient information on climb performance to be able to determine the climb gradient that can be achieved during the departure phase in the prevailing conditions.

Limitations Imposed by Weather Conditions

4.6.1 Flights to be conducted under the visual flight rules shall not be commenced unless current weather reports and forecasts indicate that visual meteorological conditions exist along the flight-planned route.

4.6.3 Flights shall not be continued towards the planned destination aerodrome unless current weather reports indicate conditions at that aerodrome, or at least one alternate destination aerodrome, are at or above specified minima.

4.6.4 Aeroplanes on approach to land shall not exceed aerodrome operating minima, except in emergency situations.

4.6.5 A flight may not be conducted in known or expected icing conditions unless the aeroplane is equipped to cope with such conditions.

Fuel and Oil Supply

4.8.1 A flight may not be commenced unless the aeroplane carries sufficient fuel and oil to complete the flight safely, considering the weather conditions and any expected delays.

In-flight Emergency Instruction

4.11 In an in-flight emergency, the pilot-in-command shall ensure that passengers and crew are instructed in appropriate emergency action.

Weather Reporting by Pilots

4.12 If weather conditions are encountered that are likely to affect the safety of other flights, they should be reported as soon as possible.

Hazardous Flight Conditions

4.13 Hazardous flight conditions encountered in-flight such as volcanic ash and dust-storms, other than those associated with weather conditions, should be reported as soon as possible.

Instruction – General

4.17 An aeroplane may be taxied on the movement area of an aerodrome only if the person at the controls:

- has been authorised by the owner, lessee or agent to do so;
- is fully competent to taxi the aeroplane;
- is qualified to use the radio if radio communications are required;
- has received instruction from a competent person in aerodrome layout, routes, signs, marking, lights, ATC signals and instructions, phraseology and procedures and is able to conform safely to the operational standards required for the safe movement of aeroplanes at the aerodrome.

Refuelling with Passengers on Board

4.18.1 ICAO recommends that aircraft should not be refuelled while passengers are boarding, on board, or leaving the aircraft, unless it is attended by the pilot-in-command or another qualified person who is able to organise an evacuation of the aircraft should it be necessary.

Aeroplane Performance and Operating Limitations

5.1 An aeroplane shall be operated:

- in compliance with its airworthiness certificate;
- within the operating limitations prescribed by the certificating authority of the State of registry.

5.2 Placards, listings and instrument markings containing operating limitations prescribed by the State of registry shall be displayed in the aeroplane.

Aeroplane Instruments and Equipment

6.1 In addition to the minimum equipment necessary to satisfy the Certificate of Airworthiness, an aeroplane shall carry the instruments, equipment and documents appropriate to the planned flight.

6.2 An aeroplane shall be equipped with instruments that will enable the flight crew to control the flightpath of the aeroplane, carry out any required procedural manoeuvre, and observe the operating limitations of the aeroplane in the expected flight conditions.

6.1.3 Aeroplanes shall be equipped with:
☐ an accessible first-aid kit;
☐ a safe portable fire extinguisher in the cockpit and in each passenger compartment if separate from the cockpit;
☐ a seat or berth for each person on board over a minimum age determined by the State of registry;
☐ a seat belt for each seat and restraining belts for each berth;
☐ the following manuals, charts and information:
 − the Flight Manual and other necessary related documents;
 − suitable aeronautical charts for the planned route and any diversions that could reasonably be anticipated;
 − procedures and visual signals for pilots-in-command of intercepted aircraft (for UK pilots, this is the CAA's *General Aviation Safety Sense Leaflet No. 11*; see also pages 27–30 of this text);
☐ spare fuses for replacement of those accessible in flight.

VFR Flights

6.2 Aeroplanes operating on VFR flights shall be equipped with:
☐ a magnetic compass;
☐ an accurate timepiece that indicates the time in hours, minutes and seconds;
☐ an altimeter;
☐ an airspeed indicator;
☐ additional instruments or equipment that may be prescribed by the appropriate authority.

Flights over Water

6.3.2 All single-engined landplanes when flying over water beyond gliding distance from land should carry one life-jacket or equivalent flotation device for each person on board, stowed in an easily accessible position for its intended user.

NOTE *Landplanes* above includes amphibious aircraft operated as landplanes.

6.3.3 All aeroplanes on extended flights over water shall be equipped as follows:

☐ When over water and more than 50 nautical miles from land suitable for an emergency landing:
   − one life-jacket or equivalent flotation device for each person on board, stowed in an easily accessible position for its intended user.

☐ When over water and more than 100 nautical miles from land suitable for an emergency landing in the case of single-engined aeroplanes, and more than 200 nautical miles, in the case of multi-engined aeroplanes capable of continuing flight with one engine inoperative:
   − live-saving rafts capable of carrying all persons on board, stowed for ready access in an emergency, provided with appropriate life-saving equipment;
   − equipment for making pyrotechnic distress signals.

Flights over Designated Land Areas

6.4 Aeroplanes flying over land areas designated by the State as being areas in which search and rescue would be especially difficult, shall be equipped with appropriate signalling devices and life-saving equipment.

*Now complete **Exercises 11 – Operation of Aircraft.***

Distress, Urgency, Safety and Warning Signals (Rule 49)

Distress Signals

If a pilot feels that the aeroplane is in grave and imminent danger and wants immediate help, then the following signals, either together or separately, before a message will alert others to the fact.

Radiotelephony (R/T or RTF): on the frequency in use or on the emergency service frequency 121.5 MHz "Mayday Mayday Mayday", followed by the message, e.g.

> **Mayday Mayday Mayday**
> *London Centre (or name of station addressed)*
> *Golf Alpha Bravo Charlie Delta*
> *Engine has failed*
> *Making a forced landing*
> *Five miles north of the field*
> *Three thousand feet and descending*
> *Heading two eight zero*
> *Student pilot*

NOTE Although not an international (ICAO) requirement, inclusion of pilot qualification in an emergency message enables the controller to plan a course of action best suited to the pilot's ability.

Also, inexperienced civil pilots are invited to use the term "TYRO" when transmitting an emergency message on a military frequency to indicate to the controller their lack of experience. This will ensure that instructions issued should be able to be followed by the pilot without any difficulty.

VISUAL SIGNALLING can be made with lights, pyrotechnics or flares to indicate **distress:**
- ☐ S-O-S in morse code: *dit-dit-dit dah-dah-dah dit-dit-dit;*
- ☐ a succession of pyrotechnic single reds fired at short intervals; or
- ☐ a red parachute flare.

SOUND SIGNALLING (other than radiotelephony) can be used to indicate **distress:**
- ☐ S-O-S in morse code: *dit-dit-dit dah-dah-dah dit-dit-dit;*
- ☐ a continuous sound, with any apparatus.

Urgency and Safety Signals

The following signals, either together or separately before a message, mean that an aircraft is in difficulties which compel it to land, but the pilot considers that he does not need immediate assistance:

☐ a succession of white pyrotechnics;

☐ repeated switching ON and OFF of landing lights;

☐ repeated switching ON and OFF of navigation lights (in a manner distinguishable from normal flashing navigation lights).

The following, either together or separately, mean that the aircraft has an **urgent** message to transmit concerning the safety of a ship, aircraft, vehicle or other property, or of a person on board or within sight of the aircraft from which the signal is given:

RADIOTELEPHONY (R/T OR RTF) on the frequency in use or on the emergency service frequency 121.5 MHz, "Pan-Pan Pan-Pan Pan-Pan", e.g.

> *Pan-Pan Pan-Pan Pan-Pan*
> *Cranfield Approach*
> *Golf Bravo Charlie Delta Echo*
> *An aircraft has force-landed two miles south of Olney*
> *Occupants appear safe and have evacuated the aircraft*

VISUAL SIGNALLING X-X-X in morse code:

> *"dah-dit-dit-dah dah-dit-dit-dah dah-dit-dit-dah"*

SOUND SIGNALLING (other than by radio): X-X-X in morse:

> *"dah-dit-dit-dah dah-dit-dit-dah dah-dit-dit-dah"*

Use of Transponder

In a radar environment, make use of your transponder to indicate an emergency (Code 7700), a radio failure (Code 7600), or unlawful interference/hijack (Code 7500).

If Mode C (altitude-reporting) is available on your transponder, it should be selected.

Danger Areas

The Danger Area Crossing Service (DACS) and Danger Area Activity Information Service (DAAIS) are available for certain UK Danger Areas. Full details are included in AIP ENR 5, and CAA 1:500,000 aeronautical charts include a list of the Danger Areas covered by each service, including the relevant Air Traffic Service Unit to contact. *Pooley's Flight Guide* also contains information on these Services.

DANGER AREA CROSSING SERVICE (DACS) is available for certain Danger Areas. The relevant areas(identified on the chart by the prefix †) and Unit Contact Frequencies to be used are shown below. For availability of the services see UK AIP RAC 5-3, column 5.

D001 .... ST MAWGAN APP 126·5MHz*
D003 & D004 .... PLYMOUTH MIL 121·25MHz or
                                                LONDON MIL VIA LONDON INFO 124·75MHz
D006 .... CULDROSE APP 134·05MHz*
D006A, D007, D007A, D007B,
D008, D008A, D008B, D009 & D009A .... PLYMOUTH MIL 121·25MHz*

DANGER AREA ACTIVITY INFORMATION SERVICE (DAAIS) is available for certain Danger Areas shown on this chart (identified by the prefix §). The Nominated Air Traffic Service Units (NATSUs) to be used are shown below. See UK AIP RAC 5-1. Pilots are advised to assume that a Danger Area is <u>active</u> if no reply is received from the appropriate NATSU.

D015 .... BOURNEMOUTH TWR 125·6MHz
✳D026 .... LONDON INFORMATION 124·75MHz
D036, D037, D038, D039, D040, D041,
D048, D049, D053, D053A, D054,
D055, D056, D057, D058 & D059 .... LONDON INFORMATION 124·75MHz/124·6MHz
                                                    /125·475MHz

■ *Figure 12-1* **Example of DACS and DAAIS information on a CAA 1:500,000 aeronautical chart**

NOTE Pilots who realise they have inadvertently entered a Prohibited Area should leave the area as quickly as possible without descending, even if no radio instruction is received.

Low Fuel Situations on Aircraft Inbound to UK

If a pilot wishes to alert ATC to the need for a priority landing because his fuel state is becoming critical, he should declare an emergency using Mayday or Pan, to ensure priority handling.

A pilot should only make such a call if he believes the aircraft to be in danger, and the seriousness of the emergency call should reflect the actual fuel state.

See AIC 9/2003 (Pink 51) for more information before planning a flight inbound to the UK.

Pilots should plan to arrive overhead their destination aerodrome with at least enough fuel to:

☐ make an approach to land;

☐ carry out a missed approach;

☐ fly to an alternate aerodrome and carry out the subsequent approach and landing; and

☐ hold at an alternate aerodrome for 45 minutes.

Note that this recommendation is a bare minimum only. It is good airmanship to carry more fuel than this to allow for unforeseen circumstances.

Now complete **Exercises 12 – Distress, Urgency, Safety and Warning Signals (Rule 49).**

Search and Rescue (SAR)

Search and Rescue in the UK

Search and Rescue in the event of a mishap is controlled by a Rescue Coordination Centre (RCC), who will act upon reports received from any source. A joint civil/military response may result. The types of service, responsible authorities and procedures are contained in AIP GEN 3-6.

A continuous listening watch is held on the aeronautical **emergency VHF frequency 121.5 megahertz,** so, if a pilot is unable to transmit a Mayday (Emergency), Pan-Pan (Urgency) or 'Lost' call on the frequency that he is already using, then a call on 121.5 MHz should be made.

Alerting Service

An Alerting Service is available for all aircraft that are known by the Air Traffic Services to be operating within the UK Flight Information Regions (FIRs). The responsibility for initiating action normally rests with the Air Traffic Services Unit (ATSU) which was last in communication with the aircraft in need of SAR assistance or which receives such information from an external source.

Autotriangulation on 121.5 MHz

Most of the United Kingdom land mass to the east and south of Manchester above 3,000 ft amsl, and down to 2,000 ft amsl in the vicinity of the London airports, is covered by a position-fixing service which operates on 121.5 MHz. The system utilises DF (direction-finding) bearing information and provides almost instantaneous aircraft position-fixing. It is activated simply by the pilot calling on the emergency frequency (121.5).

Emergency Transponder Codes

In an emergency it is a good idea to select the appropriate transponder *special purposes* code:

- Code **7700** – indicates an emergency condition (except that if already transmitting a code and receiving an air traffic service, that code will normally be retained).
- Code **7600** – indicates a radio failure.
- Code **7500** – indicates unlawful interference with the planned operation of the flight (unless circumstances warrant use of Code 7700).

Mode C (altitude reporting), if fitted, should be operated with all of the above codes.

SAR Watch Procedures

Aircraft desirous of a constant SAR watch can file a flight plan with Air Traffic Control. This is advised if:

☐ flying more than 10 nm from the coast;
☐ flying over remote or hazardous areas *(remote areas* include northern Scotland and almost all of the west coast down to Cornwall);
☐ flying an aircraft not fitted with a suitable radio.

If an aircraft is expected at an aerodrome, the pilot must inform the Air Traffic Control Unit or other Authority at that aerodrome as quickly as possible of:

☐ any change in intended destination; and/or
☐ any estimated delay in arrival of 45 minutes or more.

These requirements are contained in Rule 20(1), which is intended to apply to delays prior to departure caused by technical problems, weather deterioration, late passengers, etc., say for a flight for which a flight plan has already been submitted.

If a pilot decides whilst en route **not** to land at the planned destination, but at another airfield, then he must inform ATC at the original destination (or request that they be informed) prior to his **planned ETA plus 30 minutes** at that field, otherwise SAR action will commence.

Search and Rescue (ICAO Annex 12)

Annex 12 applies to the establishment, maintenance and operation of search and rescue services in the territories of ICAO contracting States and over the high seas, and to the coordination of SAR services between States.

SAR Service Provision

2.1 States shall provide search and rescue service within their territories 24 hours a day. This service shall be provided regardless of the nationality of an aircraft (or its occupants) in distress.

SAR Regions

2.2 States shall define regions within which they will provide search and rescue service. These regions will not overlap those of other States.

SAR Services Units

2.3 States shall establish a rescue coordination centre in each search and rescue region.

Cooperation

3.1 States shall coordinate their search and rescue organisations with those of neighbouring States.

3.2 States shall arrange for aircraft, ships and local services and facilities that are not part of the SAR organisation to help with search and rescue and to assist the survivors of aircraft accidents.

Dissemination of information

3.3 States shall publish all information necessary for the entry of rescue units of other States into their territories.

Information about Emergencies

5.1 Any authority or any part of the SAR organisation that believes an aircraft is in an emergency shall give immediately all available information to the appropriate rescue coordination centre.

Rescue coordination centres shall, on receipt of information concerning aircraft in emergency, evaluate the situation and determine the extent of action required.

When information on aircraft in danger is received from sources other than air traffic services units, the rescue coordination centre shall determine which emergency phase applies and invoke the appropriate procedures.

In addition, ICAO recommends that States should encourage anyone who observes an accident or who believes an aircraft is in danger to inform immediately the appropriate rescue coordination centre.

Procedures for Rescue Coordination Centres during Emergency Phases

5.2 UNCERTAINTY PHASE. During the uncertainty phase the rescue coordination centre shall cooperate with air traffic services units and other bodies to ensure that incoming reports are evaluated quickly.

ALERT PHASE. When the alert phase is invoked, the rescue coordination centre shall immediately alert SAR units and initiate any necessary action.

DISTRESS PHASE. When an aircraft is believed to be in distress, the rescue coordination centre shall:

- initiate action by SAR units;
- determine the position of the aircraft, estimate the degree of uncertainty of this position, and accordingly determine the extent of the area to be searched;
- notify the operator;
- notify adjacent rescue coordination centres if it seems likely that their help will be required;
- notify the appropriate air traffic services unit of information received from other sources;
- request at an early stage that aircraft, ships and coastal stations not included in the SAR organisation assist the operation by:

- maintaining a listening watch for transmissions from the aircraft in distress or from an emergency locator transmitter;
- assisting the aircraft in distress as far as practicable;
- informing the rescue coordination centre of any developments;
☐ formulate and update when necessary a plan for the SAR operation, and make that plan available to appropriate authorities;
☐ notify the State of Registry of the aircraft;
☐ notify the appropriate accident investigation authorities.

These actions shall be carried out in the above order unless circumstances dictate otherwise.

Procedures for Pilots at Accidents

5.8.1 If a pilot-in-command observes another aircraft or surface craft in distress, he shall, unless unable or considers it unnecessary:
☐ keep in sight the craft in distress for as long as necessary;
☐ if his position is not known with certainty, take steps to determine it precisely;
☐ report to the rescue coordination centre or air traffic services unit as much of the following information as possible:
- type of craft in distress, its identification and condition;
- its position, expressed in either geographical coordinates, or in distance and true bearing from a distinctive landmark or from a radio navigation aid;
- time of observation (in hours and minutes UTC);
- number of people seen;
- whether people have abandoned the craft in distress;
- number of persons afloat;
- apparent physical condition of survivors;
☐ act as instructed by the rescue coordination centre or the air traffic services unit.

5.8.1.1 If the first aircraft to reach an accident scene is not a SAR aircraft it shall take charge of on-scene activities of all other subsequently arriving aircraft until the first SAR aircraft arrives.

If, in the meantime, such aircraft is unable to establish communication with the appropriate rescue coordination centre or air traffic services unit, it shall, by mutual agreement, hand over to a capable aircraft until a SAR aircraft arrives.

5.8.2 When an aircraft has to direct a surface craft to the location of an aircraft or surface craft in distress, it shall do so by transmitting precise instructions by any means at its disposal.

Procedures for Pilots Intercepting Distress Transmissions

5.9 When the pilot-in-command of an aircraft intercepts a distress signal and/or message, he shall:

☐ record the position of the aircraft if given;

☐ if possible take a bearing on the transmission;

☐ inform the appropriate rescue coordination centre or air traffic services unit of the distress transmission, giving all available information;

☐ at his discretion, while awaiting instructions, proceed to the position given in the transmission.

Search and Rescue Signals

If an aeroplane makes a forced landing, the survivors should, if it is deemed necessary, make a call on the radio if it is still functioning, or use some or all of the following methods of attracting attention when search aircraft or surface craft are seen or heard.

1. Fire distress flares or cartridges.

2. Use some object with a bright flat surface as a heliograph to flash sunlight at the searching craft.

3. Flash a light.

4. Fly anything in the form of a flag and, if possible, make the international distress signal by flying a ball, or something resembling a ball, above or below it.

5. Blow whistles.

6. Deploy fluorescent markers to leave a trail in the sea.

7. Lay out the following ground–air visual signals (as appropriate), forming the symbols as large as possible (at least 2 or 3 m long) with materials which contrast with the background. These are the standard international signals.

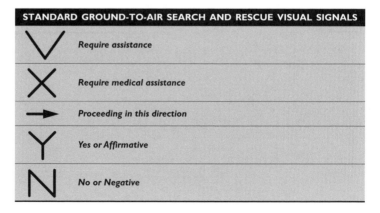

STANDARD GROUND-TO-AIR SEARCH AND RESCUE VISUAL SIGNALS

V — *Require assistance*

X — *Require medical assistance*

→ — *Proceeding in this direction*

Y — *Yes or Affirmative*

N — *No or Negative*

The following air-to-ground signals by aircraft mean that the ground signals have been understood:

☐ during daylight hours:
  – rocking the aircraft's wings;
☐ during darkness:
  – flashing on and off twice the aircraft's landing lights, or if not so equipped, switching on and off twice its navigation lights.

Lack of the above signal indicates that the ground signal is not understood.

Now complete **Exercises 13 – Search and Rescue (SAR).**

Accident Investigation Regulations

Accident Investigation in the UK

~~An accident must be notified if, between the time when anyone~~ boards an aircraft with the intention of flight, and such time as all have left it:

- anyone is killed or seriously injured while in or on the aircraft, or by direct contact with any part of the aircraft (including any part which has become detached from it) or by direct exposure to jet blast, except when the death or serious injury is from natural causes, is self-inflicted or is inflicted by other persons or is suffered by a stowaway hiding outside the areas normally available in flight to the passengers and crew; or

- the aircraft incurs damage or structural failure, other than:
  - engine failure or damage, when the damage is limited to the engine, its cowling or accessories;
  - damage limited to propellers, wingtips, aerials, tyres, brakes, fairings, small dents or punctured holes in the aircraft skin:

- unless it adversely affects its structural strength, performance or flight characteristics and which would normally require major repair or replacement of the affected component;

- the aircraft is missing or is completely inaccessible.

When a notifiable accident occurs the aircraft commander (or, if he is killed or incapacitated, the operator) must notify:

- the Chief Inspector of Air Accidents by the quickest available means; and
- the local police authority, when the accident occurs in or over the UK.

For further reading on this subject see AIC 97/2002 (Pink 43).

Purpose of Accident Investigations

The sole purpose of an accident investigation is to determine the cause and to ensure that steps are taken or recommendations are made to prevent recurrence of the same type of accident. The accident investigators have no remit to attach blame or apportion liability for an accident. It is the duty of the commander of an aircraft to assist the accident investigators in their inquiries.

The legal responsibility for implementing the safety recommendations of the investigation falls on the CAA. In order to eliminate 'blame' and to encourage reporting, a Confidential Human Factors Reporting Programme (CHIRP) is in operation.

There is a separate system for confidential reporting of general aviation accidents. The reports (GA Feedback) are available free of charge from CHIRP, Freepost (GI3439), Building F13, Room 129, DERA Farnborough, GU14 6BR (tel: Freephone 0800 214645).

Aircraft Accident and Incident Investigation (ICAO Annex 13)

The aptly numbered Annex 13 describes the procedures for ICAO States to follow during the investigation of aircraft accidents and incidents. These procedures are intended to facilitate prompt and efficient reporting of aircraft accidents by competent experts and to encourage cooperation among such experts of different States. Furthermore, States are urged to distribute accident information for the benefit of safety in air navigation worldwide. In a nutshell, the philosophy is to allow everyone to learn from the mistakes of others.

When investigating an accident, the investigator is entitled to:
☐ visit the scene of the accident;
☐ examine the wreckage;
☐ obtain witness information and suggest areas of questioning;
☐ have access to all relevant information and evidence;
☐ participate in off-scene investigations such as component examinations, tests and simulations;
☐ make submissions regarding the investigation.

Now complete
Exercises 14 – Accident Investigation Regulations.

ICAO Annex Terminology

The following summarises terms used in the ICAO Annexes pertinent to the JAR-FCL. You should be familiar with these terms because they are used in the PPL examinations.

ICAO ANNEX TERMINOLOGY

Accident

Event associated with the operation of an aircraft in which the aircraft sustains significant damage, causes significant damage, or causes personal injury. Specifically, an event that occurs between the time any person boards the aircraft with the intention of flight and the time all persons have disembarked, where:

- *a person is fatally or seriously injured as a result of:*
  - *being in the aircraft; or*
  - *being in direct contact with any part of the aircraft, including parts that have fallen off the aircraft; or*
  - *direct exposure to jet blast;*

Note: Exceptions are when the injuries are from natural causes, self-inflicted or inflicted by other persons, or when the injuries are to stowaways hiding outside the areas normally available to passengers and crew (such as cargo bays).

- *the aircraft sustains damage or structural failure which:*
  - *jeopardises the structural strength, performance or flight characteristics of the aircraft; or*
  - *would normally require major repair or replacement of the affected component;*

Note: Exceptions are engine failure or damage (when the damage is limited to the engine, its cowlings or accessories), damage limited to propellers, wing tips, antennas, tyres, brakes, fairings, small dents or puncture holes in the aircraft skin.

- *the aircraft is missing or is completely inaccessible.*

Note: An aircraft is considered to be missing when the official search has been terminated and the wreckage has not been found.

Advisory airspace

Airspace of defined dimensions, or a designated route, within which air traffic advisory service is available.

Advisory route

Designated route along which air traffic advisory service is available.

Aerial work

Aircraft operations where aircraft are used for specialised purposes, such as agriculture, construction, fish-spotting, photography, surveying, search and rescue etc.

Aerodrome

Defined area of land or water used for the arrival, departure and surface movement of aircraft.

ICAO ANNEX TERMINOLOGY

Aerodrome beacon
An aeronautical beacon used to indicate the location of an aerodrome from the air.

Aerodrome control service
Air traffic control service for aerodrome traffic.

Aerodrome control tower
A unit established to provide air traffic control services to aerodrome traffic.

Aerodrome elevation
The elevation (height above sea level) of the highest point of the landing area at the aerodrome.

Aerodrome identification sign
A sign at an aerodrome that indicates the name of the aerodrome from the air.

Aerodrome reference point
Designated geographical location of an aerodrome.

Aerodrome traffic
All traffic on the manoeuvring area of an aerodrome and all aircraft flying in the vicinity of an aerodrome.

Note: An aircraft is considered to be 'in the vicinity of an aerodrome' when it is in, entering, or leaving an aerodrome traffic circuit.

Aerodrome traffic circuit
The specified path to be flown by aircraft operating in the vicinity of an aerodrome.

Aeronautical beacon
An aeronautical ground light visible from all directions, either continuously or intermittently, to indicate

the location of a particular point on the surface of the earth.

Aeronautical fixed service (AFS)
A telecommunication service between specified fixed points provided primarily for the safety of air navigation and for the regular, efficient and economical operation of air services.

Aeronautical ground light
A light provided to aid air navigation (not a light on an aircraft).

Aeronautical Information Publication (AIP)
A document issued by a State that contains permanent aeronautical information essential to air navigation.

Aeronautical station
A land (or sea) station in the aeronautical mobile service.

Aeronautical telecommunication service
A telecommunication service provided for any aeronautical purpose.

Aeroplane
A power-driven heavier-than-air aircraft that derives its lift from aerodynamic reactions on fixed aerofoils, i.e. fixed-wing.

Airborne collision avoidance system (ACAS)
Aircraft system based on secondary surveillance radar (SSR) transponder signals which indicates to a pilot potential conflicting aircraft that are equipped with SSR transponders. ACAS operates independently of any ground-based equipment.

ICAO ANNEX TERMINOLOGY

Aircraft
Any machine that can support itself in the atmosphere, by means other than the reactions of air against the earth's surface.

Aircraft identification
A group of letters, numbers or a combination thereof which makes up the callsign of an aircraft.

Aircraft observation
A meteorological observation made from an aircraft in flight.

Aircraft proximity
A situation where minimum safe separation distances between aircraft in flight have been exceeded.

Aircraft stand
A designated area on an aerodrome apron for the parking of aircraft.

Air-ground communication
Two-way radio communication between aircraft in flight and ground (or sea) stations.

AIRMET information
Information issued by a met office about weather conditions or expected weather conditions that may affect the safety of aircraft. Such information is in addition to previously issued forecasts.

AIRPROX
Code word used in an air traffic incident report to designate aircraft proximity.

Airship
A power-driven lighter-than-air aircraft.

Air-report
A report from an aircraft in flight containing specific information on position, operation and meteorological conditions.

Air-taxiing
Movement of a helicopter above the surface of an aerodrome, normally in ground effect and at a groundspeed of less than 20 knots.

Air traffic
All aircraft in flight or operating on the manoeuvring areas of aerodromes.

Air traffic advisory service
A service provided within advisory airspace to ensure separation in so far as practical between aircraft operating on IFR flight plans.

Air traffic control clearance
Authorisation for an aircraft to proceed under conditions specified by an air traffic control unit. This term is often abbreviated to 'clearance'.

Air traffic control instruction
A directive issued by air traffic control that requires a pilot to take a specific action.

Air traffic control service
A service provided to (a) expedite the flow of air traffic and (b) to prevent collisions between aircraft in flight and on the manoeuvring area, and between aircraft and ground obstructions.

Air traffic control unit
Aerodrome control tower, area control centre or approach control office.

ICAO ANNEX TERMINOLOGY

Air traffic service
Flight information service, alerting service, air traffic advisory service or air traffic control service.

Air traffic services airspaces
Airspaces of defined dimensions, alphabetically designated (Classes A to G), within which specific types of flights may operate and for which specific air traffic services and rules of operation apply.

Air traffic services unit
Air traffic control unit, flight information centre or air traffic services reporting office.

Airway
A corridor-shaped control area equipped with radio navigation aids.

Alerting service
Service which notifies appropriate organisations of aircraft that require search and rescue aid.

Alert phase
Where concern is registered regarding the safety of an aircraft and its occupants.

Alternate aerodrome
An aerodrome to which an aircraft may proceed if it becomes either impossible or inadvisable to proceed to or land at the intended destination aerodrome. Alternate aerodromes include the following:

▪ En-route alternate: an aerodrome at which an aircraft would be able to land after experiencing an abnormal or emergency condition while en route.

▪ Destination alternate: an alternate aerodrome to which an aircraft may proceed if it becomes either impossible or inadvisable to land at the intended destination aerodrome.

Note: The departure aerodrome may also be an en-route or destination alternate aerodrome for the flight.

Altitude
The vertical distance of a point from mean sea level.

Approach control office
A unit established to provide air traffic control service to controlled flights arriving at, or departing from, one or more aerodromes.

Approach control service
Air traffic control service for arriving or departing controlled flights.

Approach sequence
The order in which two or more aircraft are cleared to approach to land at an aerodrome.

Appropriate ATS authority
The authority designated by a State as being responsible for providing air traffic services in its territory.

Appropriate authority
(a) Regarding flight over the high seas: the relevant authority of the State of Registry; (b) Regarding flight over the territory of a State: the relevant authority of the State that has sovereignty over the territory being overflown.

Apron
An area on an aerodrome where aircraft can be parked for the loading and unloading of passengers, mail or cargo, refuelling or maintenance.

ICAO ANNEX TERMINOLOGY

Area control centre
Unit which provides air traffic control service to controlled flights in control areas under its jurisdiction.

Area control service
Air traffic control service for controlled flights in control areas.

Area navigation (RNAV)
A navigation method where aircraft may operate on any flightpath within the coverage of station-referenced navigation aids or within the limits of self-contained aids, or both. Such systems avoid the need to overfly ground-based radio navigation aids.

Area navigation route
An ATS route for aircraft using area navigation.

Assignment, assign
Distribution of frequencies to stations or SSR codes to aircraft.

ATIS
Automatic terminal information service: continuous repetitive broadcast of current routine aerodrome information to arriving and departing aircraft.

ATS route
A route (airway, advisory route, arrival or departure route etc.) used as necessary for the provision of air traffic services.

Balloon
A non-power-driven lighter-than-air aircraft.

Blind transmission
A radio transmission from one station to another where the transmitter cannot hear the receiver, but believes that the transmission can be received.

Broadcast
An 'all stations' transmission of air navigation information.

Ceiling
The height above ground or water of the lowest layer of cloud below 20,000 ft covering more than half the sky.

Clearance limit
The point to which an aircraft is granted an air traffic control clearance.

Clearway
A defined rectangular area at the upwind end of a runway that is suitable for the initial climb-out of aeroplanes. Will be under the control of the aerodrome authority.

Control area
A controlled airspace extending upwards from a specified height above the earth's surface.

Controlled aerodrome
An aerodrome at which a control service to aircraft is provided.
Note: This does not necessarily imply that the aircraft is within a control zone.

Controlled airspace
An airspace of defined dimensions within which air traffic control services are provided to IFR and VFR flights.

ICAO ANNEX TERMINOLOGY

Controlled flight
Any flight subject to air traffic control clearances.

Control zone
A controlled airspace extending upwards from the earth's surface to a specified upper limit.

Cruise climb
An aeroplane cruising technique resulting in a net gain in altitude as the aeroplane mass decreases.

Cruising level
A level maintained during a significant portion of a flight.

Dangerous goods
Articles or substances that are capable of posing significant risk to health, safety or property when they are transported by air.

Declared distances
Declared distances at aerodromes are agreed by the relevant authority – in the UK this is the CAA, and the distances are published in the Aerodrome section of the AIP.

☐ **Take-off run available (TORA).** The length of runway declared available and suitable for the ground run of an aeroplane taking off.

☐ **Take-off distance available (TODA).** The length of the take-off run available plus the length of the clearway, if provided.

☐ **Accelerate-stop distance available (ASDA).** The length of the take-off run available plus the length of the stopway, if provided.

☐ **Landing distance available (LDA).** The length of runway declared available and suitable for the ground run of an aeroplane landing.

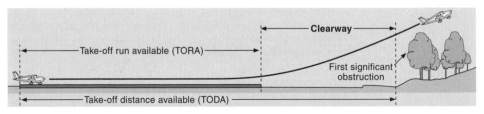

■ Figure 15-1 **TODA, TORA and clearway**

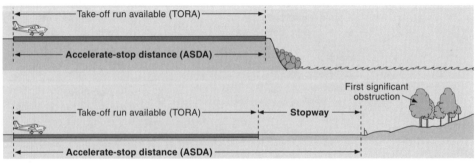

■ Figure 15-2 **Accelerate-stop distance (ASDA)**

ICAO ANNEX TERMINOLOGY

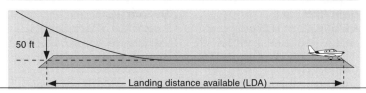

■ *Figure 15-3* **Landing distance available (LDA)**

Distress phase
Where it is reasonably certain that an aircraft and its occupants require immediate assistance or are threatened by grave or imminent danger.

Ditching
The forced landing of an aircraft on water.

Elevation
The vertical distance between a point on the earth's surface and mean sea level.

Emergency phase
Generic term meaning either uncertainty phase, alert phase or distress phase.

Estimated elapsed time
Estimated time required to proceed from one significant point to another.

Estimated off-block time
Estimated time at which the aircraft will move 'off chocks' to begin movement towards take-off.

Estimated time of arrival (ETA)
For VFR flights, the estimated time at which the aircraft will arrive over the destination aerodrome.
For IFR flights, the estimated time at which the aircraft will arrive over a point defined by radio navigation aids, from which an instrument approach procedure will begin

(if the destination aerodrome does not have an associated navigation aid, ETA is the time at which the aircraft will arrive over the aerodrome).

Expected approach time
The time at which ATC expects an arriving aircraft that has been instructed to hold will leave the holding point to complete its approach for a landing.

Filed flight plan
A flight plan as submitted to an ATS unit without subsequent changes.

Fireproof material
A material capable of withstanding heat as well as, or better than, steel

Flight crew member
A licensed crew member charged with duties essential to the operation of an aircraft during flight.

Flight Information centre
A unit that provides flight information service and alerting service.

Flight information region
An airspace of defined dimensions within which flight information service and alerting service is provided.

Flight information service
Service which provides advice and information useful to the safe and efficient conduct of flights.

ICAO ANNEX TERMINOLOGY

Flight level

A surface of constant atmospheric pressure which is related to a specific pressure datum, 1013.2 mb (hPa), and is separated from other such surfaces by specific pressure intervals. Flight levels are expressed in hundreds of feet, e.g. FL180 = 18,000 ft.

Flight Manual

A manual, associated with the Certificate of Airworthiness, containing limitations within which the aircraft is to be considered airworthy, and instructions and information necessary to the pilot for the safe operation of the aircraft.
An aircraft Flight Manual is written by the manufacturer (e.g. Piper), approved by the State of Manufacture (in the US approval is given by the FAA) and supplemented if necessary by the State of Registration (for instance, by the UK CAA). The Flight Manual forms part of the Certificate of Airworthiness.

Flight plan

Specified information provided to air traffic services units about an intended flight or portion of a flight.

Flight time

The total time from the beginning of the take-off roll until the moment the aircraft stops at the end of a flight.

Note: This definition of flight time is synonymous with the terms 'block to block' or 'chock to chock'.

Flight visibility

Visibility forward from the cockpit of an aircraft in flight.

Forecast

A description of the expected weather conditions over a specified period of time for a particular area.

General aviation operation

An aircraft operation other than a commercial air transport flight or an aerial work operation.

Glider

A non-power-driven heavier-than-air aircraft that derives its lift from aerodynamic reactions on fixed aerofoils.

Ground visibility

Visibility at an aerodrome, as reported by a meteorological observer.

Gyroplane

A power-driven heavier-than-air rotorcraft that derives its lift from aerodynamic reactions on a freely-rotating rotor in the vertical axis. Powered by a propeller on the longitudinal axis.

Hazard beacon

An aeronautical beacon used to indicate a danger to air navigation.

Heading

The direction in which an aircraft is pointing, usually expressed in degrees from north (either true, magnetic or compass).

Height

The vertical distance between a point and a specified datum (such as sea level or ground level).

Heavier-than-air aircraft

An aircraft that derives its lift mainly from aerodynamic forces.

ICAO ANNEX TERMINOLOGY

Helicopter
A heavier-than-air rotorcraft that derives its lift and control from one or more power-driven rotors on substantially vertical axes.

Heliport
An aerodrome or a defined area on a structure for the landing, taking off and surface movement of helicopters.

Holding bay
A defined area at an aerodrome where aircraft can be held or bypassed without disrupting the flow of other traffic.

Holding point
A specified location, around which an aircraft flies a standard pattern until cleared to proceed with the flight.

Identification beacon
An aeronautical beacon that flashes a coded signal such that its location can be identified.

IFR flight
A flight made under the Instrument Flight Rules.

IMC
Instrument Meteorological Conditions.

Incident
An occurrence, other than an accident, which affects or could affect the safety of an aircraft operation.

Instrument Meteorological Conditions (IMC)
Meteorological conditions expressed in terms of visibility, distance from cloud, and ceiling, less than the minima specified for visual meteorological conditions.

Investigation
A process conducted for the purpose of accident investigation which includes the gathering and analysis of information, the drawing of conclusions, including the determination of causes and, when appropriate, the making of safety recommendations.

Landing area
The area on an aerodrome used for the landing or take-off of aircraft.

Level
A generic term relating to the vertical position of an aircraft in flight – referring to either height, altitude or flight level.

Lighter-than-air aircraft
An aircraft that is supported in flight mainly by its buoyancy in the air (e.g. a hot-air balloon).

Location indicator
A four-letter code assigned to the location of an aeronautical fixed station (could be either an aerodrome or a met station).

Manoeuvring area
The part of an aerodrome used for the taxiing, take-off and landing of aircraft, excluding aprons.

Marker
An object displayed above ground level to indicate an obstacle or boundary, e.g. the orange-and-white striped wedge-shaped markers that delineate an aerodrome boundary.

ICAO ANNEX TERMINOLOGY

Marking
A symbol or group of symbols displayed on the surface of the movement area to convey aeronautical information, e.g. the double white cross used to indicate gliding is in progress.

Meteorological information
Meteorological report, analysis, forecast or other statement relating to existing or expected weather conditions.

Meteorological office
An office that provides a meteorological service for international air navigation.

Meteorological report
A statement of observed weather conditions at a specific place at a specific time.

Mode (SSR)
Mode of operation of SSR transponder, e.g. Mode C (altitude reporting) or Mode A.

Movement area
The part of an aerodrome used for the taxiing, take-off and landing of aircraft, including the manoeuvring area and apron(s).

Night
The hours between the end of evening civil twilight and the beginning of morning civil twilight or such other period between sunset and sunrise as may be prescribed by the appropriate aviation authority.

Non-instrument runway
A runway for the use of aircraft using visual approach procedures only.

Non-radar separation
Separation distances between aircraft when position information is obtained from sources other than radar.

NOTAM
Notice to Airmen, which contains urgent information concerning the establishment, condition or change in any aeronautical facility, service, procedure or hazard.

Obstacle
Any fixed or mobile object (or part thereof) located on the surface movement area of an aerodrome that extends above a defined height

Operator
A person, organisation or enterprise engaged in or offering facilities in aircraft operations.

Pilot-in-command
The pilot responsible for the operation and safety of an aircraft during flight time.

Pressure altitude
An atmospheric pressure expressed in terms of altitude which corresponds to that pressure in the standard atmosphere (i.e. 1013.2 mb set in altimeter subscale).

Primary radar
Radar system that uses reflected radio signals.

Primary surveillance radar
Radar surveillance system that uses reflected radio signals.

ICAO ANNEX TERMINOLOGY

RADAR
Radio detection system which provides information on range, position and elevation of objects.

Radar approach
An approach to land where the final approach phase is directed by a radar controller.

Radar clutter
Unwanted signals displayed on a radar screen, caused by interference, static etc.

Radar contact
When the radar position of a particular aircraft is seen and identified on a radar display.

Radar control
Where radar information is used directly in the provision of air traffic control.

Radar controller
An air traffic controller qualified to use radar information.

Radar display
Electronic display (screen, monitor) which uses radar information to depict the position and movement of aircraft.

Radar identification
When the position of a particular aircraft is seen on a radar display and positively identified by the air traffic controller.

Radar monitoring
Use of radar to provide aircraft with information on their deviations from planned flightpath and deviations from air traffic control clearances.

Radar separation
Separation distances used when aircraft position information is provided by radar sources.

Radar service
A service provided by means of radar.

Radar unit
Part of an air traffic services unit that uses radar.

Radar vectoring
Where a radar controller issues heading instructions to aircraft, based on radar information.

Radio direction-finding station
A radio station that determines the relative direction of other transmitting stations.

Radiotelephony
A form of radio communication used mainly for the exchange of speech information.

Reporting point
A geographic location at which the position of an aircraft in flight can be reported.

Rescue coordination centre
Unit responsible for organising search and rescue operations within a certain area.

Rescue unit
A group of people trained and equipped to perform search and rescue operations.

Rotorcraft
A power-driven heavier-than-air aircraft that is supported in flight by reactions of air on one or more rotors.

ICAO ANNEX TERMINOLOGY

Runway
A defined rectangular area on an aerodrome used for the take-off and landing of aircraft.

Runway guard lights
A light system which alerts pilots or vehicle drivers that they are about to enter an active runway.

Runway visual range (RVR)
The distance along which the pilot of an aircraft on the centre-line of a runway can see the runway surface markings or lights.

Safety recommendation
A proposal made by the accident investigation authority of the State conducting an investigation, based on information derived from the investigation, with the intention of preventing accidents or incidents.

Search and rescue aircraft
An aircraft equipped to conduct search and rescue missions.

Search and rescue region
An area of defined dimensions within which search and rescue service is provided.

Search and rescue services unit
A generic term meaning either rescue coordination centre, rescue subcentre or alerting post.

Secondary radar
Radar system where an 'interrogating' radio signal transmitted from the radar station prompts a 'reply' signal to be sent from an aircraft transponder.

Secondary surveillance radar (SSR)
Radar system that uses transmitters/receivers (interrogators) and transponders.

Serious incident
An event that almost resulted in an accident.

Note: The only difference between an accident and an incident is the result: damage and/or injury = accident; could have been damage and/or injury = incident.

Serious injury
An injury sustained by a person in an accident which:
- requires hospitalisation for more than 48 hours (from within 7 days of the accident);
- results in a bone fracture (apart from simple fractures of fingers, toes or nose);
- involves lacerations which cause severe haemorrhage, nerve, muscle or tendon damage;
- involves injury to any internal organ;
- involves second or third degree burns, or any burns that affect more than 5% of body surface;
- involves verified exposure to infectious substances or harmful radiation.

SIGMET information
Information issued by a met office concerning weather conditions or expected weather conditions that may affect the safety of flights.

Signal area
An area on an aerodrome used for the display of ground signals.

ICAO ANNEX TERMINOLOGY

Slush

Water-saturated snow which, with a heel-and-toe slap-down motion against the ground, will be displaced with a splatter.

Snow (on the ground)

▣ **Dry snow.** Snow which can be blown if loose or, if compacted by hand, will fall apart again on release.

▣ **Wet snow.** Snow which, if compacted by hand, will stick together and tend to form a snowball.

▣ **Compacted snow.** Snow which has been compressed into a solid mass that resists further compression and will hold together or break into lumps if picked up.

Special VFR flight

A VFR flight cleared by air traffic control to operate within a control zone in meteorological conditions below VMC.

State of Design

The State that has jurisdiction over the organisation responsible for the design of a particular aircraft type.

State of Manufacture

The State that has jurisdiction over the organisation responsible for final assembly of an aircraft.

State of Occurrence

The State in which an aircraft accident or incident occurs.

State of the Operator

The State in which the operator's principal place of business is located, or if there is no such place, the operator's permanent residence.

State of Registry

The State (nation) in which an aircraft is registered.

Stopway

A defined rectangular area on the ground at the end of the take-off end of a runway, prepared as a suitable area in which an aeroplane can stop in the case of an abandoned take-off.

Surveillance radar

Radar equipment used to determine the range and position of aircraft in azimuth.

Take-off runway

A runway intended for take-off only.

Taxi-holding position

A designated position at an aerodrome where taxiing aircraft may be required to hold before entering or crossing a runway.

Taxiing

Movement on the surface of an aerodrome of an aircraft under its own power, excluding take-off and landing.

Taxiway

A defined path on an aerodrome for the taxiing of aircraft.

Terminal control area

A control area normally established around a major aerodrome.

Threshold

The beginning of the usable portion of a runway (normally indicated by 'piano key' markings).

ICAO ANNEX TERMINOLOGY

Touchdown zone
The portion of a runway, beyond the threshold, where it is intended landing aeroplanes first contact the runway.

Track
The path of an aircraft in flight over the earth's surface.

Traffic avoidance advice
Advice given by air traffic control to pilots to assist in collision avoidance.

Traffic information
Information given by air traffic control to pilots regarding other known traffic near or on the flight-planned route.

Transition altitude
The altitude at or below which the vertical position of aircraft is controlled by reference to altitudes (i.e. with Regional QNH set). Transition altitudes vary considerably between countries: 3,000 ft in the UK, 18,000 ft in the USA.

Transition layer
Airspace between the transition altitude and transition level.

Transition level
The lowest flight level available for use above the transition altitude.

Uncertainty phase
When the safety of an aircraft and its occupants is uncertain.

VFR flight
Flight conducted under the Visual Flight Rules.

Visibility
The distance over which prominent unlighted objects by day and prominent lighted objects by night can be seen.

Visual approach
An approach to land by an IFR flight where part or all of an instrument approach procedure is not completed and the approach is conducted by visual reference to terrain.

Visual meteorological conditions
Meteorological conditions expressed in terms of visibility, distance from cloud, and ceiling, equal or better than specified minima.

VMC
Visual meteorological conditions.

Waypoint
A specific geographical location used by an area navigation system.

Now complete **Exercises 15 – Annex Terminology**

Section **Two**

Meteorology

The Atmosphere

The solid earth is surrounded by a mixture of gases which are held to it by the force of gravity. This mixture of gases we know as **air** and the space it occupies around the earth we call the **atmosphere**. The atmosphere is of particular importance to pilots because it is the medium in which aeroplanes fly.

Air Density

Air density decreases with altitude.

The force of gravity exists between each individual air molecule and the earth. This has the effect of drawing them together, causing the air molecules to crowd around the surface of the earth and squeeze closer together than at altitude. Therefore, there are more molecules per unit volume of air at sea level than at higher altitudes. For example, at 40,000 ft altitude, the number of molecules in a cubic metre of air is only about half that at sea level.

Density is the mass per unit volume and, on average, the density of air at sea level is 1,225 grammes per cubic metre.

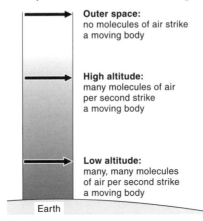

Figure 16-1 **Air density decreases with increasing altitude**

Three reasons why air density is so important to pilots are that:
- **the lift force** supporting the aeroplane's weight is generated by the flow of air around the wings;
- **engine power** is generated by burning fuel and air; and
- **we need to breathe air** in order to live.

In air that is dense:
- **the required lift** can be generated at a lower true airspeed (V);
- **greater engine power** is available;
- **a person's breathing is easier** and more oxygen is taken into the lungs.

The Atmosphere

The earth spins on its axis, carrying the atmosphere with it and tending to throw the air to the outside. For this reason, the atmosphere extends further into space above the equator than above the poles.

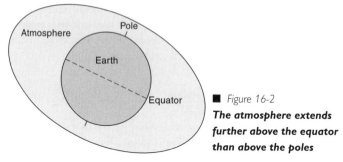

■ *Figure 16-2*
The atmosphere extends further above the equator than above the poles

The surface temperature at the equator is greater than at the poles. This also contributes to the approximate oval shape of the atmosphere (the atmosphere is thickest where the surface temperature is highest). As the hottest part of the surface moves north or south of the equator with the seasons, the relative position of the thickest part of the atmosphere moves accordingly.

Subdivision of the Atmosphere

The atmosphere is divided vertically into four regions – the troposphere, the stratosphere, the mesosphere and the thermosphere. Light aircraft flying occurs in the troposphere, although high flying jets may cruise in the stratosphere. The boundary between the two regions is known as the **tropopause.**

The tropopause occurs at a height of about 20,000 ft over the poles, and at about 60,000 ft over the tropics. In the 'average' International Standard Atmosphere it is assumed to occur at 36,000 ft. Most 'weather' occurs in the troposphere.

There are some significant differences between the stratosphere and the troposphere:

☐ Temperature falls with height gained in the troposphere, but is constant in the stratosphere at about –57°C (in general terms).

☐ There is marked vertical movement of air in the troposphere, with warm air rising and cool air descending, on both a large and a small scale.

☐ The troposphere contains almost all of the water vapour in the atmosphere and so cloud formation rarely extends beyond the tropopause (although occasionally large cumulonimbus clouds with strong and fast vertical development may push into the stratosphere).

Composition of Air

Air is a mixture of gases, as shown in the table below.

COMPOSITION OF AIR	
Gas	**Volume (%)**
Nitrogen	78%
Oxygen	21%
Other gases (argon, carbon dioxide, neon, helium, etc.)	1%
Total	100%

Air always contains some **water vapour.** This is most important because it is water vapour that condenses to form clouds, from which we get the precipitation (rain, snow, hail, etc.) that is essential to life on earth.

Maritime and Continental Air Masses

Maritime air is more moist than continental air.

Air over an ocean (known as maritime air) will absorb moisture from the body of water and will, in general, contain more water vapour than the air over a continent (known as continental air), particularly if the land mass consists largely of desert areas.

In other words, a maritime air mass is more moist than a continental air mass, and so an air mass moving in across the UK from over the Atlantic Ocean is likely to carry more moisture than an air mass whose origin is continental Europe and Asia.

While nitrogen is the main constituent of air, the other constituents vital to aviation are:

☐ oxygen (to support life and combustion);
☐ water vapour (to produce weather).

Humidity

The water molecule (H_2O) is a relatively light molecule and its presence in large numbers in an air mass lowers its density, which affects both the aerodynamic performance of an aeroplane and the power production from the engine, making performance slightly poorer on a damp day compared with a dry day. This is a consideration when operating in conditions of **high relative humidity.**

Another consideration in relation to operating in conditions of high relative humidity is the possibility of ice forming in the carburettor as the air is cooled by expansion while mixing with the vaporising fuel.

*Now complete **Exercises 16 – The Atmosphere.***

Heating Effects in the Atmosphere

The Sun

The sun radiates energy and heats the earth.

The source of energy on earth is the sun, which radiates electro-magnetic energy (light, radio waves, etc.). We experience this solar radiation as **heat** and **light**.

The wavelength of solar radiation are such that a large percentage penetrates the earth's atmosphere and is absorbed by the earth's surface, causing its temperature to increase. The ground in turn heats that part of the atmosphere which is very close to it, with the result that any parcel of air warmer than the surrounding air will then rise.

Seasonal Variations

The earth orbits around the sun once in every year and, because the earth's axis is tilted, this gives rise to the four seasons. The solar radiation received at a place on earth is more intense in summer than in winter, when its surface is presented to the sun at a more oblique angle.

Note that although the earth is closest to the sun in March and September, summer occurs when the earth is at its furthest from the sun. Therefore the distance from the earth to the sun is not a significant factor in the amount of radiation received, in comparison with the angle at which the rays are presented.

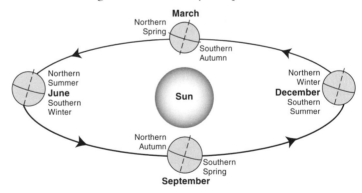

■ *Figure 17-1 **Solar radiation received at the earth's surface is more intense in summer***

Solar Heating

Heating from solar radiation is greatest in the tropics.

Solar radiation is like a torch beam that produces more intense light on a perpendicular surface than on an oblique surface. Since solar radiation strikes tropical regions from directly overhead, or almost directly overhead, right throughout the year, the heating is quite intense.

In contrast, the sun's rays strike polar regions of the earth at an oblique angle and, during the winter situation (the northern summer is shown in the diagram), they may not strike the polar regions at all.

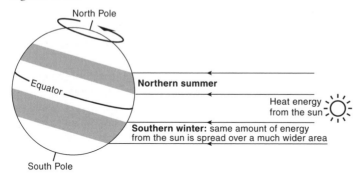

■ Figure 17-2 **Surface heating is greatest in the tropics and least at high latitudes**

General Circulation

The greater heating of the earth's surface in the tropics causes the air in contact with it to become relatively warm. This leads to the air expanding, becoming less dense in the process and thus commencing to rise, which in turn causes it to spread out in the upper regions of the atmosphere. As a result, new air moves in across the earth's surface to replace the air which has risen.

In contrast, the cooler air over the polar regions sinks down, creating a large-scale vertical circulation pattern in the troposphere. This process is known as the **general circulation** pattern and consists of three main 'cells':

☐ the polar cell ;
☐ the mid–latitude cell (or Ferrel cell); and
☐ the tropical cell (or Hadley cell).

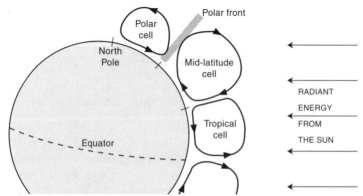

■ Figure 17-3 **The general circulation pattern**

The hot and less dense air rising over the tropics creates a band of low pressure at the earth's surface known as the **equatorial trough**, into which other surface air will move (known as *convergence*).

The cool and dense air subsiding in the polar regions creates a high-pressure area at the earth's surface in the very high latitudes and the surface air will spread outwards (known as *divergence*).

Terrestrial Re-radiation

Heat energy in the earth's surface is re-radiated into the atmosphere but, because its wavelength is longer than solar radiation, it is more readily absorbed in the atmosphere, especially by water vapour and carbon dioxide. It is this absorption of heat from the earth that is the main process which causes weather.

In summary:

- **solar radiation** penetrates the atmosphere and heats the earth's surface; then
- **the earth re-radiates this energy** and heats the lower levels of the atmosphere.

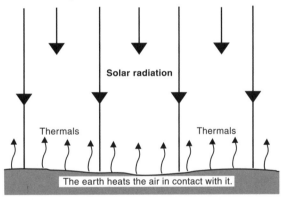

■ *Figure 17-4* **Indirect heating of the atmosphere by the sun**

Local Heating and Cooling

The earth rotates on its axis once every 24 hours, causing the apparent motion of the sun across the sky, which results in day and night on the earth.

Solar heating of the earth's surface occurs only by day, but terrestrial re-radiation of heat energy occurs continually through both the day and the night. The net result is that the earth's surface heats up by day, reaches its maximum temperature about mid-afternoon and cools by night, reaching its minimum temperature around sunrise.

This continual heating and cooling on a daily basis is called the **diurnal variation of temperature** – a typical daily pattern of heating and cooling that is most extreme in desert areas and more moderate over the oceans.

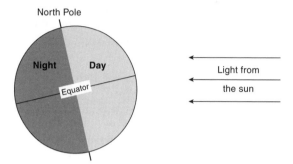

■ *Figure 17-5* **The earth's rotation causes day and night**

Surface Heating

The heating of various surfaces and the temperatures which they reach depends upon a number of factors:

Different surfaces heat differently.

1. **The specific heat of the surface.** Water requires more heat energy to raise its temperature by 1°C than land, therefore land will heat more quickly (and also cool more quickly). Compared with the sea, land is warmer by day and cooler by night. Scientifically we say water has a higher specific heat than land.

2. **The reflectivity of the surface.** If the solar radiation is reflected from the surface it can, obviously, *not* be absorbed. Snow and water surfaces have a high reflectivity and so will not be heated as greatly as absorbent areas such as a ploughed field or a dense jungle.

3. **The conductivity of the surface.** Ocean currents transfer heat through water motion, causing the sea to be heated to a greater depth than a land surface.

Cloud Cover

Cloud coverage by day stops some of the solar radiation penetrating to the earth's surface, resulting in reduced heating of the earth and lower temperatures. The air in contact with the surface will therefore be subject to less heating by day.

Cloud cover affects surface heating and cooling.

By night, however, cloud coverage causes the opposite effect and prevents some of the heat energy escaping from the earth's surface. The atmosphere beneath the cloud will experience less cooling and higher temperatures will result.

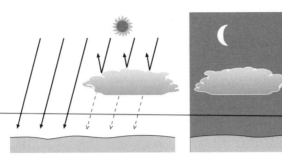

■ *Figure 17-6* **Cloud reduces surface heating by day and cooling by night**

The Transfer of Heat Energy

Heat energy may be transmitted from one body to another, or redistributed in the one body by a number of means, including:

1. RADIATION. All bodies transmit energy in the form of electro-magnetic radiation, the higher the temperature of the body, the shorter the wavelength of the radiation. Radiation from the sun is therefore of shorter wavelength than the re-radiation from the earth, which is much cooler.

2. ABSORPTION. Any body in the path of radiation will absorb some of its energy. How much it absorbs depends upon both the nature of the body and of the radiation. A densely forested area will absorb more solar radiation than snow–covered mountains.

3. CONDUCTION. Heat energy may be passed within the one body, or from one body to another in contact with it, by conduc-tion. Iron is a good conductor of heat – wood is a poor conductor, as is air. A parcel of air in contact with the earth's surface is heated by conduction, but will not transfer this heat energy to neigh-bouring parcels of air. This is a very significant factor in the production of weather systems.

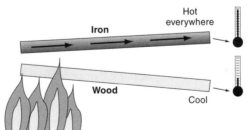

■ *Figure 17-7* **Some things are good conductors of heat; others are not**

4. CONVECTION. A body in motion carries its heat energy with it. A mass of air that is heated at the earth's surface will expand, becoming less dense, and rise. As it rises, it will carry its heat energy higher into the atmosphere – which is the process called convection.

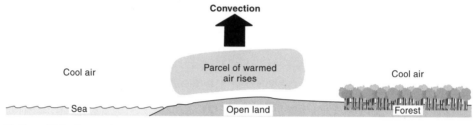

■ *Figure 17-8* **Convection**

5. ADVECTION. When air moves in to replace air that has risen by convection, this horizontal motion of air is known as advection. The air mass, moving horizontally by advection, will of course bring its heat energy (and moisture content) with it.

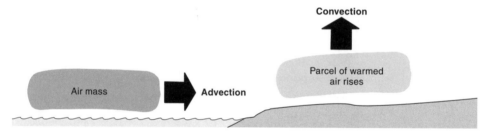

■ *Figure 17-9* **Advection is the horizontal transfer of air and heat energy**

The general vertical circulation pattern of airflow that occurs on a large scale around the earth also happens on a smaller scale in localised areas.

Local Air Movements

The Sea Breeze by Day

The process known as a sea breeze occurs on sunny days, when the land heats more quickly than the sea. The air above the land becomes warmer and rises (usually by mid- to late afternoon). This sets up the situation for a vertical circulation pattern with the cooler air from offshore moving in to replace the air from over the land which has risen due to surface heating.

The vertical extent of a sea breeze is usually only 1,000 or 2,000 ft.

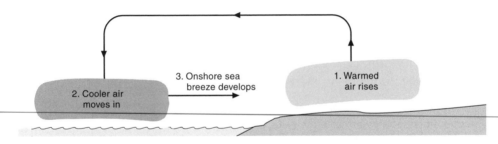

■ Figure 17-10 **The sea breeze – a small circulation cell**

Sea breezes may have a significant effect on aerodromes near a coastline. If the sea breeze opposes the general wind pattern it is quite possible that the wind velocity at circuit height will be quite different from that at ground level. Windshear and some turbulence may be experienced as the aeroplane passes from one body of air to the other.

The cooler air moving in over the land as a sea breeze may bring fog or mist with it, causing visibility problems. This could be the situation if a sea fog exists and, as the day progresses and the land heats up, an onshore sea breeze develops, moving the fog in across the land.

The Land Breeze by Night

By night, the land cools more quickly than the sea, causing the air above it to cool and subside. The air over the sea is warmer and will rise.

A land breeze could hold a sea fog offshore early in the day but, as the land warms, the land breeze could die out and a sea breeze develop, bringing the sea fog in and possibly causing visibility problems at coastal aerodromes.

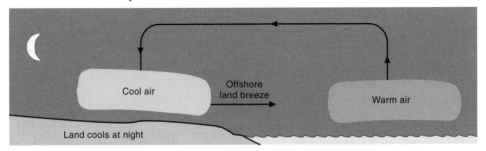

■ Figure 17-11 **The land breeze blows offshore at night**

Katabatic Winds

During night-time the earth's surface loses a lot of heat through terrestrial radiation and cools down. This is particularly the case on clear, cloudless nights. The air in contact with the ground loses heat to it by conduction, cools down, becomes denser and starts to sink.

In mountainous regions the cool air will flow down the sides of the mountain slopes and into the valleys, creating what is called a **katabatic wind**. In certain areas, katabatic winds of 30 knots or so can be flowing down the slopes of large mountains into the valleys by sunrise.

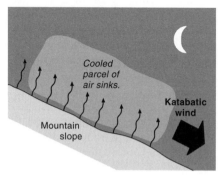

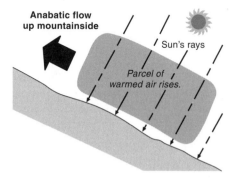

■ *Figure 17-12* **Katabatic winds blow down mountain slopes and valleys at night**

■ *Figure 17-13* **Anabatic winds drift up mountain slopes by day**

Anabatic Winds

Heating of a mountain slope by day causes the air mass in contact with it to warm, decreasing its density and causing it to flow up the slope. Since its flow uphill is opposed by gravity, the daytime anabatic wind is generally a weaker flow than the night-time katabatic wind.

Temperature Inversions

The general pattern of temperature distribution in the atmosphere is that **temperature decreases with height**. In the International Standard Atmosphere (ISA), which is a purely theoretical model of the atmosphere used as a measuring stick, the temperature is assumed to fall by 2°C for each 1,000 ft climbed in the stationary air mass.

On clear nights when the earth loses a great deal of heat by terrestrial radiation and cools down, the air in contact with its surface also cools by conduction. The cooler air tends to sink and not mix with air at the higher levels. This will lead to the air at ground level being cooler than the air at altitude, a phenomenon known as a **temperature inversion**.

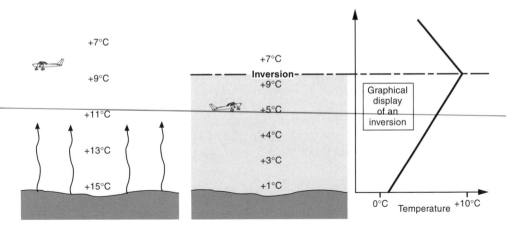

■ *Figure 17-14* **The normal temperature situation and the temperature inversion**

The temperature inversion may exist only for twenty feet or a few hundred feet, but it is important as it can have byproducts such as ground fog or windshear. As a pilot passes through the inversion layer he may experience a slight 'bump'.

Temperature Measurement

Heat Energy

As a body of matter absorbs heat energy its molecules become more agitated. This agitation and motion is measured as temperature, which can be used as a measure of heat energy. The temperature at which molecular agitation would theoretically be zero is known as absolute zero, i.e. zero on the scientific absolute scale of temperature, and this occurs at $-273°C$.

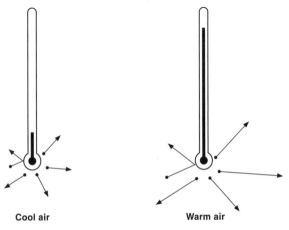

■ *Figure 17-15* **Air molecules 'move' more in warm air**

Temperature Scales

Various scales are used to measure temperature, but the one used in most countries for aviation and meteorology is the **Celsius** scale (previously called *centigrade*). The Celsius scale divides the temperature difference between the boiling and freezing points of water into 100 degrees. On the Celsius scale, water boils at 100°C and freezes at 0°C.

In the United States, the Fahrenheit temperature scale is predominantly used, though the Celsius scale is being introduced. The Fahrenheit scale is based on water boiling at 212°F and freezing at 32°F.

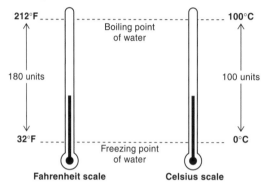

■ *Figure 17-16* **Different temperature scales measure the same thing**

There is a requirement for pilots (who may fly in different countries) to be able to convert from °F to °C and vice versa. Flight navigation computers have conversion scales marked on them to make these conversions easy, but you should still know the mathematical relationship that exists between °C and °F.

☐ Temperature in °F = ⅗ × °C + 32.
☐ Temperature in °C = ⅝ × (°F − 32).

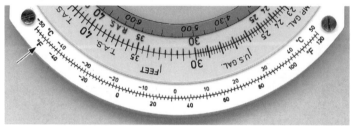

■ *Figure 17-17* **Temperature conversion scale on a navigation computer**

Now complete

Exercises 17 – Heating Effects in the Atmosphere.

Atmospheric Pressure

The molecules that make up the air move at high speed in random directions and bounce off any surface that they encounter. The force which they exert on a unit area of that surface is called the **atmospheric pressure.**

Because of the fewer air molecules at the higher levels and the less weight of molecules pressing down from above, the atmospheric pressure decreases with height. An aircraft flying at altitude, or a town located in a mountainous area, will therefore experience a lower pressure than at sea level.

Atmospheric pressure can be measured with either:

- A **mercury barometer,** where atmospheric pressure at sea level can support a column of approximately 30 inches of mercury by pushing it into a partial vacuum.
- An **aneroid barometer,** where a flexible metal chamber that is partially evacuated is compressed by the atmospheric pressure. The aneroid is used in aircraft altimeters to measure the atmospheric pressure and equate variations in air pressure to changes in altitude.

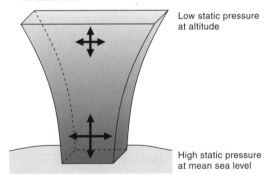

Low static pressure
at altitude

High static pressure
at mean sea level

■ *Figure 18-1* **Atmospheric pressure decreases with height**

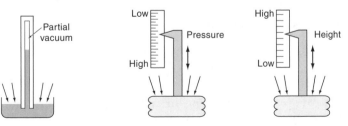

■ *Figure 18-2* **The mercury barometer, aneroid barometer and altimeter**

In the United Kingdom the unit of pressure used in aviation is the **millibar** (abbreviated **mb**). This is gradually being changed throughout the world to an equivalent unit known as the **hecto-pascal**, abbreviated **hPa**; however, for the foreseeable future **milli-bars are to be used in the UK**. At sea level on a standard day, the atmospheric pressure is 1013.2 mb (or hPa).

As height above sea level is gained in the lower levels of the atmosphere, the pressure drops by approximately 1 mb per 30 ft. An altimeter that reads 0 ft at sea level is calibrated to read 30 ft when the pressure has dropped by 1 mb, 300 ft when it has dropped 10 mb, and 1,000 ft when it has dropped 33 mb, and so on.

Altimetry is covered in more detail in other parts of this series – *Altimeter Setting Procedures* in this volume, *Vertical Navigation* in Volume 3, and *Pressure Instruments* in Volume 4.

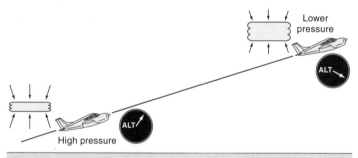

■ *Figure 18-3* **The altimeter equates a pressure drop with an increase in height**

The atmospheric pressure at a particular place is continually varying. These variations may be:

☐ **irregular,** due to pressure systems which are passing, intensify-ing or weakening; and

☐ **regular,** due to the daily heating and cooling effects of the sun – known as the **semi-diurnal variation of pressure**, since it has a 12 hour (½ day) cycle, with maximum pressures occur-ring at about 10 a.m. and 10 p.m., and pressures some 2 or 3 mb less occurring in between.

The Pressure Gradient

Readings of atmospheric pressure are taken at many locations on a regular basis and, because these locations are at various altitudes, the readings are reduced to a sea level value for comparison purposes.

Places that are experiencing the same calculated sea level pres-sures are then joined with lines on maps. These lines are known as **isobars**. For reasons of clarity on meteorological charts, the isobars are usually spaced at 2 mb intervals or greater. (Isobars on the weather maps published in daily newspapers, which cover a large area, are spaced at 8 mb intervals.)

The isobars often form patterns on the weather map that are very meaningful – some isobars surrounding areas of high pressure, others surrounding areas of low pressure, and others being straight.

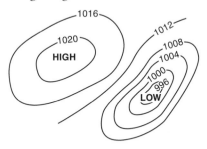

■ *Figure 18-4* **Isobars join places of equal sea level pressure**

The variation of pressure with horizontal distance is called the **pressure gradient**, which occurs at right-angles to the isobars. If the isobars are very close together rapid changes in pressure occur and the pressure gradient is said to be 'steep' or 'strong'; if they are widely spaced, the pressure changes are more gradual and so the pressure gradient is 'flat' or 'weak'.

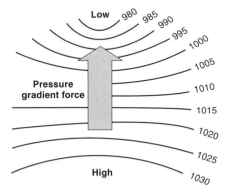

■ *Figure 18-5* **Pressure gradient**

There is a natural tendency for air to flow from areas of high pressure to areas of low pressure. This increases with the steepness of the pressure gradient. As we shall see in the chapter on 'Wind', the final airflow is not directly from high to low, but is somewhat modified as a result of the earth's rotation.

Regional QNH

Flights cruising with QNH set in the altimeter subscale should reset the QNH, as advised, to the **current Regional QNH** (also known as Regional Pressure Setting), and also when entering a new 'Region'. If the QNH is **not** revised, then if flying towards an area of low pressure, the aeroplane will gradually descend, assuming that the pilot is maintaining a constant indicated altitude as indicated on his altimeter.

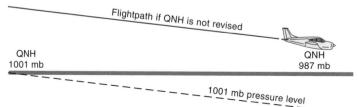

Flightpath if QNH is not revised

QNH
1001 mb

QNH
987 mb

1001 mb pressure level

■ *Figure 18-6* **When cruising, reset the altimeter to the current Regional QNH**

This situation could be dangerous, since the altimeter will be indicating the height above the pressure level in the subscale, and this will be higher than the aeroplane's actual height above sea level, i.e. the altimeter will over-read unless the subscale is reset to the lower QNH.

Conversely, flying towards an area of high pressure the situation is reversed. The altimeter will under-read, the aeroplane being at a higher altitude than that indicated, unless the subscale is period-ically reset to the higher QNH.

In certain circumstances is it considered to be good airmanship to use alternatives to the Regional QNH. For instance, pilots flying beneath a Terminal Control Area (TMA) may set the QNH pertaining to that TMA, e.g. the London QNH when under the London TMA. Alternatively, pilots flying within 25 nm of an aerodrome may set the QNH pertaining to that aerodrome.

Now complete **Exercises 18 – Atmospheric Pressure.**

The International Standard Atmosphere (ISA)

To act as a measuring stick against which to compare the actual atmosphere existing at any place and time, an **International Standard Atmosphere** has been defined. It is commonly known as the 'ISA' (pronounced *"eye-sah"*) or the ICAO Standard Atmosphere. (ICAO is the International Civil Aviation Organisation, the aviation 'wing' of the United Nations, see Chapter 1).

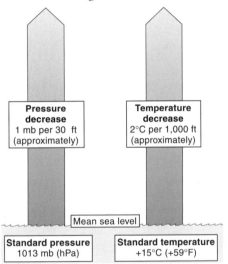

Pressure decrease
1 mb per 30 ft
(approximately)

Temperature decrease
2°C per 1,000 ft
(approximately)

Mean sea level

Standard pressure
1013 mb (hPa)

Standard temperature
+15°C (+59°F)

■ *Figure 19-1* ***The International Standard Atmosphere (ISA)***

The International Standard Atmosphere is based on mean sea level values of:
- pressure 1013.2 mb(hPa);
- temperature +15°C;
- density 1,225 gm/cubic metre;
- each of which decreases as height is gained.

The temperature of the atmosphere falls at a rate of 2°C per 1,000 ft until the tropopause is reached at about 36,000 ft amsl, above which the temperature is assumed to remain constant at −57°C. (In precise terms, the temperature lapse rate is 1.98°C per 1,000 ft and remains a constant −56.5°C at 36,090 ft).

The pressure falls at about 1 mb(hPa) per 30 ft gained in the lower levels of the atmosphere (up to about 5,000 ft.).

The ISA and the 'Real' Atmosphere

The actual atmosphere can differ from the ISA in many ways. Sea level pressure varies from day to day, indeed from hour to hour, and the temperature also fluctuates between wide extremes at all levels.

The variation of ambient pressure throughout the atmosphere – both horizontally and vertically – is of great significance to pilots, as it affects the operation of the altimeter. This is considered elsewhere in this series.

Now complete

Exercises 19 – The International Standard Atmosphere (ISA).

Wind

What is Wind?

The term *wind* refers to the flow of air over the earth's surface. This flow is almost completely horizontal, with only about ¹⁄₁₀₀₀ of the total flow being vertical.

Despite being only a small proportion of the overall flow of air in the atmosphere, vertical airflow is extremely important to weather and to aviation, since it leads to the formation of cumuliform clouds and thunderstorms. Some vertical winds are so strong, in fact, that they are a hazard to aviation and can destroy aircraft. In general, however, the term *wind* is used in reference to the horizontal flow of air.

How Wind Is Described

Both the direction and strength of a wind are significant and are expressed thus:

☐ **wind direction** is the direction from which the wind is blowing and is expressed in degrees measured clockwise from north;

☐ **wind strength** is expressed in knots (abbreviated kt).

Direction and strength together describe the **wind velocity,** which is usually written in the form 270/35, or 270/35KT (to specify the unit), i.e. a wind blowing from due west (270°) at a strength of 35 knots.

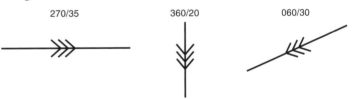

■ *Figure 20-1* **Examples of wind velocity**

Meteorologists relate wind direction to **true north,** and so all winds that appear on forecasts are expressed in degrees true (°T). For example **34012KT** on an aerodrome forecast or meteorological observation means a wind strength of 12 knots from a direction of 340°T. **23020G35KT** indicates a wind from 230°T at 20 knots, with **gusts** to 35 knots.

Runways, however, are described in terms of their **magnetic direction,** so that when an aeroplane is lined up on the runway ready for take-off, its magnetic compass and the runway direction should agree, at least approximately.

The wind direction relative to the runway direction is extremely important when taking off and landing. For this reason, winds passed to the pilot by the tower have direction expressed in **degrees magnetic**. This is also the case for the recorded messages on the Automatic Terminal Information Service (ATIS) that pilots can listen to on the radio at some airfields.

Veering and Backing

A wind whose direction is changing in a clockwise direction is called a **veering** wind. For example, following a change from 150/25 to 220/30, the wind is said to have veered.

A wind whose direction is changing in an anticlockwise direction is called a **backing** wind. A change from 100/15 to 030/12 is an example of a wind that has backed.

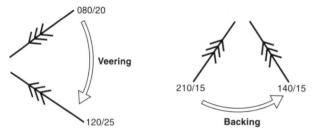

■ Figure 20-2 **A veering wind and a backing wind**

What Causes a Wind to Blow?

A change in velocity (speed and/or direction) is called acceleration. Acceleration is caused by a force (or forces) being exerted on an object, be it an aeroplane, a car or a parcel of air.

The combined effect of all the forces acting on a body is known as the net (or resultant) force, and determines the acceleration of the body. If all of the forces acting on a parcel of air balance each other so that the resultant force is zero, then the parcel of air will not accelerate, but will continue to move in a straight line at a constant speed (or stay still). A steady wind velocity is known as **balanced flow**.

The Pressure Gradient Force

In the atmosphere the force that is usually responsible for starting a parcel of air moving is the **pressure gradient force**. This acts to move air from areas of **high** pressure to areas of **low** pressure.

Places on the earth where air pressure is the same are joined on meteorological maps with lines which are called **isobars**, so the pressure gradient force will act at right angles to these and in the direction from the high to the low pressure. The stronger the pressure gradient (i.e. the greater the pressure difference over a given distance), the greater this force will be and, consequently, the stronger the wind will blow.

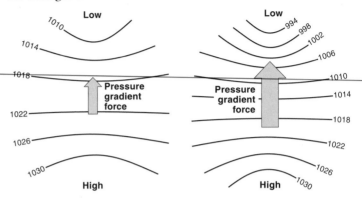

■ *Figure 20-3* **The pressure gradient force will start a parcel of air moving**

If the pressure gradient force was the only force acting on a parcel of air, it would continue to accelerate towards the low pressure, getting faster and faster, and eventually the high- and low-pressure areas would disappear because of this transfer of air.

We know that this, in fact, does not occur, so there must be some other force that acts on the parcel of air to prevent it rushing from the high-pressure area into the low-pressure area. This other force is due to the rotation of the earth and is known as the **Coriolis** force.

The Coriolis Force

The Coriolis force acts on a **moving** parcel of air. It is not a 'real' force, but an apparent force resulting from the passage of the air over the rotating earth.

Imagine a parcel of air that is stationary over Point A on the equator. It is in fact moving with Point A as the earth rotates on its axis from west to east. Now, suppose that a pressure gradient exists with a high pressure at A and a low pressure at Point B, directly north of A. The parcel of air at A starts moving towards B, but still with its motion towards the east due to the earth's rotation.

The further north one goes from the equator, the less is this easterly motion of the earth and so the earth will lag behind the easterly motion of the parcel of air. Point B will only have moved to B', but the parcel of air will have moved to A". In other words, to an observer standing on the earth's surface the parcel of air will **appear** to turn to the right. This effect is due to the Coriolis force.

If the parcel of air was being accelerated in a southerly direction from a high-pressure area in the northern hemisphere towards a low near the equator, the earth's rotation towards the east would 'get away from it' and so the airflow would appear to turn right also – A having moved to A', but the airflow having only reached B" to the west.

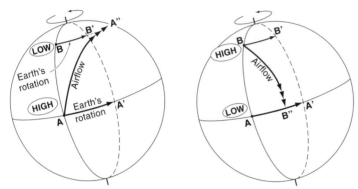

■ Figure 20-4 **The Coriolis force acts towards the right in the northern hemisphere**

The faster the airflow, the greater the Coriolis effect – if there is no air flow, then there is no Coriolis effect. The Coriolis effect is also greater in regions away from the equator and towards the poles, where changes in latitude cause more significant changes in the speed at which each point is moving towards the east.

In the northern hemisphere, the Coriolis force deflects the wind to the **right;** in the southern hemisphere, the situation is reversed and it deflects the wind to the left.

The Geostrophic Wind

The two forces acting on a moving airstream are:
- the pressure gradient force; and
- the Coriolis force.

The pressure gradient force gets the air moving and the Coriolis effect turns it to the right. This curving of the airflow over the earth's surface will continue until the pressure gradient force is balanced by the Coriolis force, resulting in a wind flow that is steady and blowing in a direction **parallel to the isobars.** This **balanced flow** is called the **geostrophic wind.**

The **geostrophic wind** is important to a weather forecaster because it flows in a **direction** parallel to the isobars with the low pressure to its left, and at a strength directly proportional to the spacing of the isobars (i.e. proportional to the pressure gradient).

The closeness of the isobars on a chart enables a reasonable forecast of the wind strength to be made.

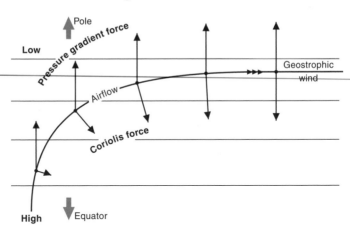

■ *Figure 20-5* **Balanced flow will occur parallel to the isobars – the geostrophic wind**

Buys Ballot's Law

Buys Ballot was a Dutchman who noticed that:

"If you stand with your back to the wind in the northern hemisphere, the low pressure will be on your left."

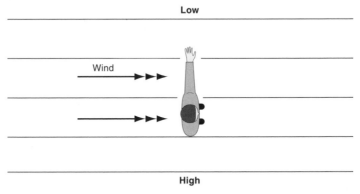

■ *Figure 20-6* **Buys Ballot's law – with your back to the wind, the low pressure is on your left (northern hemisphere)**

Flying from high to low

Flying from high to low, beware below.

If an aeroplane in the northern hemisphere is experiencing right (starboard) drift, the wind is from the left and therefore, according to Buys Ballot's law, the aeroplane is flying towards an area of lower pressure. Low pressure often has poor weather associated with it, such as low cloud, rain and poor visibility.

With right drift, the aeroplane is flying towards an area of lower pressure. Unless the pilot periodically resets the lower Regional QNHs, the altimeter will over-read, which is not a healthy situation, so beware below (see Figure 18-6).

Flying from low to high

If an aeroplane in the northern hemisphere is experiencing left (port) drift, the wind is from the right and therefore, according to Buys Ballot's law, the aeroplane is flying towards an area of higher pressure. High pressure often indicates a more stable atmosphere and generally better weather (although fog may occur).

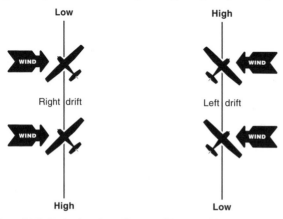

■ Figure 20-7 **Right (starboard) drift – low pressure area ahead**

■ Figure 20-8 **Left (port) drift – high pressure area ahead**

The Gradient Wind

Isobars (the lines joining places of equal pressure) are usually curved. For the wind to flow parallel to these isobars, it must be accelerated, in the sense that its direction is changing. In the same manner as a stone when being swung on a string is pulled into the turn by a force, a curving air flow must have a resultant (or net) force acting on it to pull it into the turn.

For a wind that is blowing (anticlockwise) around a **low** in the northern hemisphere, the net force results from the pressure gradient force being greater than the Coriolis force, thereby pulling the air flow in towards the **low**.

For a wind that is blowing (clockwise) around a **high** the net force results from the Coriolis force being greater than the pressure gradient force. Since the Coriolis force increases with wind speed, it follows that wind speed around a **high** will be greater than wind speed around a **low** with similarly spaced isobars.

In the northern hemisphere, the result is a wind flowing parallel to the isobars, clockwise around a **high** (known as anticyclonic motion) and anticlockwise around a **low** (known as cyclonic motion). Balanced wind flow around curved isobars is called the **gradient wind**.

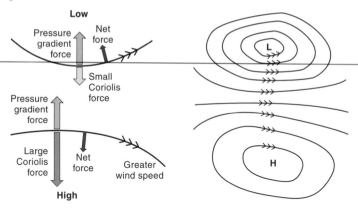

■ *Figure 20-9* **Wind flows clockwise around a high and anticlockwise around a low in the northern hemisphere**

The Surface Wind

The surface wind is very important to pilots because of the effect it has on take-offs and landings. Surface wind is measured at 10 metres (30 ft) above level and open ground, i.e. where windsocks and other wind indicators are generally placed.

Wind is usually less strong near the surface than at higher levels. The gradient wind at height that flows parallel to the curved isobars is slowed down by the friction that exists between the lower layers of the airflow and the earth's surface. The Coriolis effect will be less due to the lower wind speed, and so the wind will back in direction. The rougher the surface is, the greater the slowing down. Friction forces will be least over oceans and flat desert areas, and greatest over hilly or city areas with many obstructions.

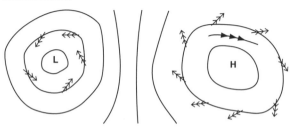

■ *Figure 20-10* **Friction causes the surface wind to weaken in strength and back in direction**

A reduced wind speed results in a reduced Coriolis force (since it depends on speed). Therefore the pressure gradient force will have a more pronounced effect in the lower levels, causing the wind to flow in towards the low-pressure area and out towards the high-pressure area rather than flow parallel to the isobars. In other words, **the surface wind tends to back compared to the gradient wind.**

Over oceans, the surface wind may slow to about two-thirds of the gradient wind strength and the backing may only be about 10°, but over land surfaces it may slow to only one-third of the gradient wind strength and be some 30° back from the gradient flow parallel to the isobars.

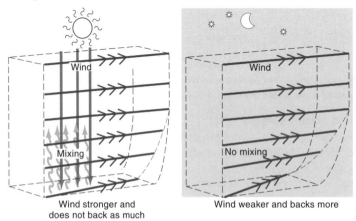

Wind stronger and does not back as much   Wind weaker and backs more

■ *Figure 20-11* **The diurnal variation of wind**

Frictional forces due to the earth's surface decrease rapidly with height and are almost negligible above 2,000 ft above ground level (agl). The turbulence due to wind flow over rough ground also fades out at about the same height above the surface.

Diurnal (Daily) Variation in the Surface Wind

During the day, heating of the earth's surface by the rays of the sun, and the consequent heating of the air in contact with it, will cause vertical motion in the lower levels of the atmosphere. This promotes mixing of the various layers of air, and consequently the effect of the gradient wind at altitude will be brought closer to the earth's surface.

The surface wind by day will resemble the gradient wind more closely than the surface wind by night, i.e. the day surface wind will be seen as a stronger wind that has veered compared to the night surface wind.

Day – veer and increase.
Night – slack and back.

During the night, mixing of the layers will decrease. The gradient wind will continue to blow at altitude, but its effects will not be mixed with the airflow at the surface to such an extent as during the day. The night wind at surface level will drop in strength and the Coriolis effect will weaken, i.e. compared to the day wind, the night wind will drop in strength and back in direction.

Localised Friction Effects

The surface wind may bear no resemblance to the gradient wind at 2,000 ft agl and above if it has to blow over and around obstacles such as hills, trees, buildings, etc. The wind will form turbulent eddies, the size of which will depend on both the size of the obstructions and the wind strength.

■ Figure 20-12 **Friction and obstacles affect the surface wind**

Flight in Turbulence

Some degree of turbulence is almost always present in the atmosphere and pilots quickly become accustomed to its minor forms. Moderate or severe turbulence, however, is uncomfortable and can even overstress the aeroplane.

Vertical gusts will increase the angle of attack, causing an increase in the lift generated at that particular airspeed and therefore an increased load factor. Of course, if the angle of attack is increased beyond the critical angle, the wing will stall and this can occur at a speed well above the published 1g stalling speed.

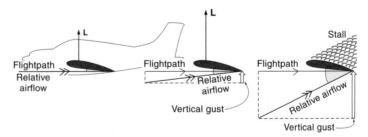

■ Figure 20-13 **Vertical gusts increase angle of attack and will increase the load factor and/or stall the wing**

Load factor (or g-force) is a measure of the stress on the aeroplane and each category of aeroplane is built to take only certain load factors. It is important that these load factors are not exceeded. One means of achieving this is to fly the aeroplane at turbulence penetration speed, which is usually 10–20% slower than normal cruise speed, but not so slow as to allow the aeroplane to stall, remembering that in turbulence the aeroplane may stall at a speed higher than that published.

When encountering turbulence:

☐ Fasten seatbelts.

☐ Hold the attitude for the desired flight phase (i.e. climb, cruise, descent), using whatever aileron movements are needed to retain lateral control, but being fairly gentle on the elevator to avoid over-stressing the airframe structurally through large changes in angle of attack and lift produced.

☐ The airspeed indicator will probably be fluctuating, so will be less useful than normally. Aim to have the airspeed fluctuate around the selected turbulence penetration speed, which may require reduced power. Use power to maintain speed.

It is of course better to avoid turbulence, and to some extent this is possible:

☐ Avoid flying underneath, in or near thunderstorms where airflows can be enormous.

☐ Avoid flying under large cumulus clouds because of the large updrafts that cause them.

☐ Avoid flying in the lee of hills when strong winds are blowing, since they will tumble over the ridges and possibly be quite turbulent as well as flowing down into valleys at a rate which your aeroplane may not be able to outclimb.

☐ Avoid flying at a low level over rough ground when strong winds are blowing.

Windshear

Windshear is the variation of wind speed and/or direction from place to place. It affects the flightpath and airspeed of an aeroplane and can be a hazard to aviation. Flight considerations involving windshear are covered in the *Windshear* chapter of Volume 4 of *The Air Pilot's Manual*.

Windshear is generally present to some extent as an aeroplane approaches the ground for a landing, because of the different speed and direction of the surface wind compared to the wind at altitude. Low-level windshear can be quite marked at night or in the early morning when there is little mixing of the lower layers, for instance when a temperature inversion exists.

Windshear can also be expected when a sea breeze or a land breeze is blowing, or when in the vicinity of a thunderstorm. Cumulonimbus clouds have enormous updrafts and downdrafts associated with them, and the effects can be felt up to 10 or 20 nautical miles away from the actual cloud. Windshear and turbulence associated with a thunderstorm can destroy an aeroplane.

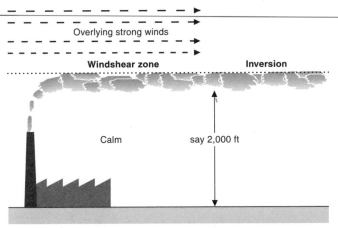

■ *Figure 20-14* **Windshear caused by a temperature inversion**

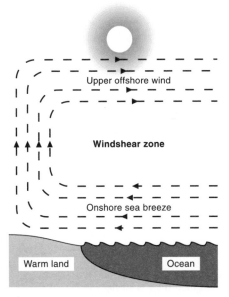

■ *Figure 20-15* **Windshear caused by a sea breeze**

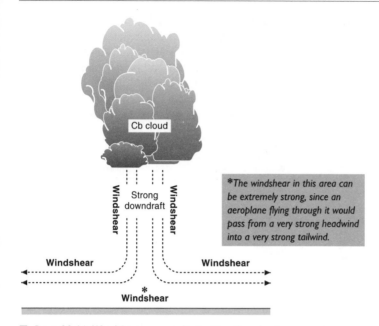

■ Figure 20-16 **Windshear near a dissipating thunderstorm can be extremely hazardous**

Wind Associated with Mountains

Wind that flows over a mountain and down the lee side can be a hazard to aviation, not only because it may be turbulent, but also because the aeroplane will have to 'climb' into it just to maintain altitude. For this reason, an aeroplane should maintain a vertical clearance of several thousand feet above mountainous areas in strong wind conditions.

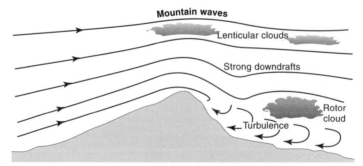

■ Figure 20-17 **Avoid flying near mountains in strong winds**

There may also be local wind effects, such as the katabatic wind that flows down cool slopes at night and in the morning, and also valley winds.

Large mountains or mountain ranges cause an effect on the wind that may extend well above ground level, resulting in **mountain waves**, possibly with associated lenticular clouds. The up-currents and down-currents associated with mountain waves can be quite strong, and can extend for 30 or 40 nm downwind of the mountains. Rotor areas may form beneath the crests of the nearer lee waves, and are often characterised by a roll cloud. There may be severe turbulence in the rotor zone.

Wind in the Tropics

In tropical areas, pressure gradients are generally fairly weak and so will not cause the air to flow at high speeds. Local effects, such as land and sea breezes, may have a stronger influence than the pressure gradient.

The Coriolis force that causes the air to flow parallel to the isobars is very weak in the tropics since the distance from the earth's axis remains fairly constant. The pressure gradient force, even though relatively weak, will dominate and the air will tend to flow more from the high-pressure areas to the low-pressure areas than parallel to the isobars.

Instead of using isobars (which join places of equal pressure) on tropical weather charts, it is more common to use:

▪ **streamlines** to indicate wind direction, which will be *outdrafts* from high-pressure areas and *indrafts* to low-pressure areas; in combination with

▪ **isotachs**, which are dotted lines joining places of equal wind strength.

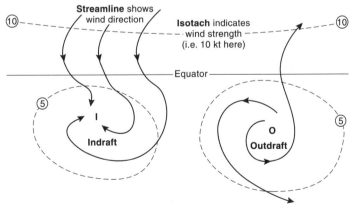

▪ *Figure 20-18* **Streamline/isotach analysis chart**

Now complete **Exercises 20 – Wind.**

Cloud

A cloud is a visible aggregate of minute particles of water and/or ice in the free air. The effect of cloud on aviation, particularly visual flight, makes them an important topic in pilot training. The presence of low *stratus* clouds, for example, can cause a flight to divert or even turn back without reaching the destination.

In 'unstable' atmospheric conditions, large build-ups of cloud can develop into thunderstorms – known in meteorological terms as *cumulonimbus*. Thunderstorms are extremely hazardous to aviation.

Unfortunately, the classification of cloud types and of individual clouds is not straightforward, because clouds may take on numerous different forms, many of which continually change. It is important to have an understanding of cloud classification because the meteorological forecasts and reports use this system to give a picture of the weather for the pilot.

The Naming of Clouds

The four main groups of clouds are:

- **cirriform** (or fibrous);
- **cumuliform** (or heaped);
- **stratiform** (or layered); and
- **nimbus** (or rain-bearing).

Clouds are further divided according to the level of their bases above mean sea level, resulting in ten basic types.

High-Level Cloud

High-level cloud has a base above 20,000 ft and looks quite fine and spidery because it is distant from the observer. As it is in a quite cold region of the atmosphere it is composed of ice crystals rather than water particles.

1. CIRRUS (Ci). Detached clouds in the form of white delicate filaments, white patches or narrow bands. These clouds have a fibrous or silky appearance.

2. CIRROCUMULUS (Cc). A thin, white patch, sheet or layer of cloud without shading, composed of very small elements in the form of grain or ripples, joined together or separate, and more or less regularly arranged. Most of the elements have an apparent width of less than 1° of arc (approximately the width of the little finger at arm's length). *Cirrus* indicates high and *cumulus* indicates lumpy or heaped.

3. CIRROSTRATUS (Cs). A transparent whitish veil of fibrous or smooth appearance, totally or partly covering the sky and generally producing a halo phenomenon (a luminous white ring around the sun or moon with a faint red fringe on the inside). *Cirrus* indicates high and *stratus* indicates layer.

Middle-Level Cloud

Middle-level cloud has a base above about 6,500 ft amsl.

4. ALTOCUMULUS (Ac). A layer of patches of cloud composed of laminae or rather flattened globular masses, rolls, etc., the smallest elements having a width of between 1° and 5° of arc (the width of three fingers at arm's length). They are arranged in groups or lines or waves which may be joined to form a continuous layer or appear in broken patches and shaded either white or grey. **Coronae** (one or more coloured rings around the sun or moon) are characteristic of this cloud. In an unstable atmosphere, the vertical development of *Ac* may be sufficient to produce precipitation in the form of **virga** (rain that does not reach the ground) or slight showers. *Alto* means middle-level and *cumulus* means heaped.

5. ALTOSTRATUS (As). A greyish or bluish cloud sheet of fibrous or uniform appearance totally or partly covering the sky and having parts thin enough to reveal the sun at least vaguely as though through ground glass. Precipitation in the form of rain or snow can occur with *As*. *Stratus* means layer, so *Altostratus* is middle-level layer cloud.

Low-Level Cloud

Low cloud has a base below about 6,500 ft amsl.

6. NIMBOSTRATUS (Ns). A dark grey cloud layer generally covering the whole sky and thick enough throughout to hide the sun or moon. The base appears diffuse due to more or less continuously falling rain or snow. At times *Ns* may be confused with *As* since it is like thick altostratus, but its darker grey colour and lack of a distinct lower surface distinguish it as *Ns*. It may be middle-level or low-level cloud. *Nimbus* means rain-bearing and *stratus* means layer.

7. STRATOCUMULUS (Sc). A grey or whitish patch or sheet of cloud which has dark parts composed of rounded masses or rolls which may be joined or show breaks between the thicker areas. Most of the rounded masses have an apparent width of more than 5° of arc (the width of three fingers at arm's length). The associated weather, if any, is very light rain, drizzle or snow. *Stratus* means layer and *cumulus* means heaped.

8. STRATUS (St). A generally grey cloud layer with a fairly uniform base, which may give precipitation in the form of drizzle. When the sun is visible through the cloud, its outline is clearly discernible. *Stratus* means layer.

9. CUMULUS (Cu). Detached clouds, generally dense and with sharp outlines, developing vertically in the form of rising mounds, domes or towers, of which the upper part often resembles a cauliflower. The sunlit parts of these clouds are mostly brilliant white. Their base is relatively dark, since sunlight may not reach it, and nearly horizontal. Precipitation in the form of rain or snow showers may occur with large cumulus. *Cumulus* means heaped.

10. CUMULONIMBUS (Cb). A heavy and dense cloud with considerable vertical extent in the form of a mountain or huge tower. At least part of its upper portion is usually fibrous or striated, often appearing as an anvil or vast plume. The base of the cloud appears dark and stormy. Low ragged clouds are frequently observed below the base and generally other varieties of low cloud, such as *Cu* and *Sc*, are joined to or in close proximity to the *Cb*.

Lightning, thunder and hail are characteristic of this type of cloud, while associated weather may be moderate to heavy showers of rain, snow or hail. *Cumulo* means heaped and *nimbus* means rain-bearing.

Noting any precipitation can help in recognising the particular type of cloud. Showers (which start and stop suddenly and may be followed by a clear sky) fall only from convective clouds such as cumulus and cumulonimbus.

Precipitation

Intermittent or continuous precipitation (which usually starts and finishes gradually, perhaps over a long period) is usually associated with stratiform cloud, e.g. drizzle from stratus and stratocumulus, heavy continuous rain or snow from nimbostratus, rain from altostratus. Showers are associated with cumuliform clouds.

The above are the ten main cloud classifications, but there are certain variations that you may see mentioned, such as:

- **Stratus fractus** and **cumulus fractus** – stratus or cumulus, as appropriate, observed as shreds or fragments below the base of nimbostratus or altostratus.
- **Castellanus** – a number of small, cumuliform clouds sharing a common base and indicating the growth of middle-level clouds in an unstable atmosphere.
- **Lenticularis** – lens-shaped clouds formed in standing waves over mountains caused by strong winds aloft and often associated with cumuliform cloud.

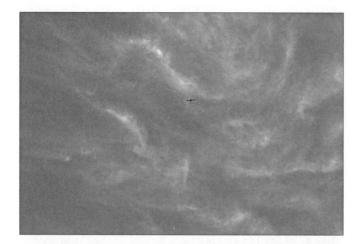

■ Figure 21-1
Filaments of cirrus (Ci)

■ Figure 21-2
Cirrostratus (Cs)

■ Figure 21-3
Altostratus (As), thickening towards horizon; Cu fractus below

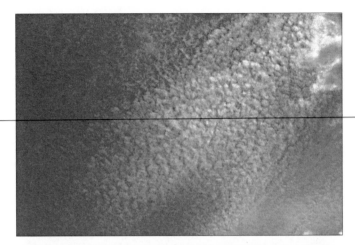

■ Figure 21-4
Altocumulus (Ac)

■ Figure 21-5
Nimbostratus (Ns)

■ Figure 21-6
Stratocumulus (Sc)

■ *Figure 21-7*
**Stratus (St); terrain
(and TV mast!) in
cloud**

■ *Figure 21-8*
Cumulus (Cu)

■ *Figure 21-9*
**Large cumulus
build-up (showers and
poor visibility below)**

■ *Figure 21-10*
Mature cumulo-nimbus (Cb) (avoid Cb clouds, with or without anvil)

■ *Figure 21-11*
Ac castellanus formation

■ *Figure 21-12*
Lenticular altocumulus

Moisture in the Atmosphere

Cloud is formed from water vapour that is contained in the atmosphere. It is taken up into the atmosphere by evaporation from the oceans and other bodies where water is present.

The Three States of Water

Water in its vapour state is not visible, but when the water vapour condenses to form water droplets we see it as cloud, fog, mist, rain or dew. Frozen water is also visible as cloud (high-level), snow, hail, ice or frost. Water therefore exists in three states – gas (vapour), liquid (water) and solid (ice).

Under certain conditions water can change from one state to the other, absorbing heat energy if it moves to a higher-energy state (from ice to water to vapour) and giving off heat energy if it moves to a lower-energy state (vapour to water to ice). This heat energy is known as **latent heat** and is a vital part of any change of state.

The three states of water, the names of the various transfer processes and the absorption or giving off of latent heat are shown in Figure 21-13.

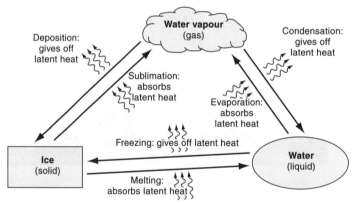

■ Figure 21-13 **The three states of water**

Humidity

The **amount of water vapour** present in the air is called **humidity**, but the actual amount is not as important as whether the air can support that water vapour or not.

Relative Humidity

When a parcel of air is supporting as much **water vapour** as it can, it is said to be **saturated** and have a **relative humidity of 100%**. If it is supporting less than its full capacity of water vapour, it is said to be unsaturated and its relative humidity will be less than 100%. Air supporting only a third of the water vapour that it could has a relative humidity of 33%. There are of course many degrees of saturation ranging from 0% to 100%. In cloud and fog it is 100%, whereas over a desert it may be 20%.

Relative humidity (RH) is defined as the ratio of water vapour actually in the parcel of air relative to what it can hold (i.e. when it is saturated) at a particular temperature and pressure.

$$RH = \frac{mass\ of\ water\ vapour}{mass\ of\ water\ vapour\ at\ saturation} \times \frac{100}{1}\ \%$$

How much water a particular parcel of air can support is dependent on the air temperature – warm air being able to support more than cool air. If the temperature falls, the amount of water vapour that the air can support decreases and so the relative humidity will increase. In other words, even though no moisture has been added, the relative humidity of a parcel of air will increase as its temperature drops.

Dewpoint Temperature

Dewpoint is the temperature at which a parcel of air becomes saturated if it cools (at constant pressure), i.e. the temperature at which it is no longer able to support all of the water vapour that it contains; the more moisture in the air, the higher its dewpoint temperature.

A parcel of air that has a temperature higher than its dewpoint will be unsaturated, i.e. its relative humidity is *less* than 100%. The closer the actual temperature is to the dewpoint, the closer the air is to being saturated.

At its dewpoint, the air will be fully saturated (RH = 100%) and if it becomes cooler than its dewpoint, then the excess water vapour will condense as visible water droplets (or, if it is cool enough, as ice). This process can be seen both when moist air cools at night to form fog, dew or frost, and as air rises and cools, the water vapour condensing into the small water droplets that form clouds. If the air is unable to support these water droplets (for example, if they become too large and heavy), then they fall as precipitation (rain, hail or snow).

Adiabatic Processes

An adiabatic process is one in which heat is neither added to nor removed from the system. The expansion and compression of gases are adiabatic processes where, although heat is neither added nor removed, the temperature of the system may change, e.g. placing your finger over the outlet of a bicycle pump will illustrate that compressing air increases its temperature.

Conversely, air that is compressed and stored at room temperature will feel cool if released to the atmosphere and allowed to expand. Reducing pressure will lower the temperature.

Cloud Formation

A very common adiabatic process that involves the expansion of a gas and its cooling is when a parcel of air rises in the atmosphere. This can be initiated by the heating of the parcel of air, causing it to expand and become less dense than the surrounding air, so that it will rise. A parcel of air can also be forced aloft as it blows over a mountain range.

Unsaturated air will cool adiabatically at about 3°C/1,000 ft as it rises. This is known as the **dry adiabatic lapse rate (DALR)**.

Cooler air can support less water vapour, so, as the parcel of air rises and cools, its relative humidity will increase. At the height where its temperature is reduced to the dewpoint temperature (i.e. relative humidity reaches 100%), water will start to condense and form cloud.

Above this height the now-saturated air will continue to cool as it rises but, because latent heat will be released as the water vapour condenses into the lower-energy liquid state, the cooling will not be as great. The rate at which saturated air cools as it rises is known as the **saturated adiabatic lapse rate (SALR)** and may be assumed to have a value of approximately half the DALR, i.e. 1.5°C/1,000 ft.

Which Cloud Type Forms?

The nature and extent of any cloud that forms depends on the nature of the surrounding atmosphere through which the parcel of air is ascending. As long as the parcel of air is warmer than its surroundings, it will continue to rise. An atmosphere in which a parcel of air, when given vertical movement, continues to move away from its original level is called **unstable**. Cumuliform clouds (i.e. heaped) may form to a high level in such an atmospheric situation; the more moisture in the air, the higher its dewpoint temperature.

The type of cloud that forms depends on the stability of the atmosphere.

If the surrounding atmosphere is warmer than the parcel of air, it will stop rising because its density will be greater than the surroundings. An atmosphere in which air tends to remain at the one level is called a **stable** atmosphere.

The stability of the atmosphere depends on the ambient lapse rate.

The rate of temperature change in the surrounding atmosphere is called the **environmental lapse rate (ELR)** and its relationship to DALR and SALR is a main factor in determining the levels of the bases and tops of the clouds that form.

The International Standard Atmosphere (ISA), which is simply a theoretical 'measuring stick' against which the actual atmosphere at any time or place can be compared, assumes an ELR of 2°C/1,000 ft. The actual ELR in a real atmosphere, however, may differ greatly from this.

Cloud Formed by Convection due to Heating

Suppose that a parcel of air overlying a large ploughed field is heated to +17°C, whereas the air in the surrounding environment is only 12°C. The heated parcel of air will start to rise, due to its lower density, and cool at the dry adiabatic lapse rate of 3°C/1,000 ft.

If the environmental lapse rate happens to be 1°C/1,000 ft, then the environmental air through which the heated parcel is rising will cool at only 1°C/1,000 ft.

Suppose that the moisture content of the parcel of air is such that the dewpoint temperature is 11°C. By 2,000 ft agl the rising parcel of air will have reached this and so water will start to condense and form cloud. At 2,000 ft agl, the environmental air will have cooled to 10°C, so the parcel of air will continue rising since it is still warmer (11°).

As the parcel of air continues to rise above the level at which cloud first forms, latent heat will be released as more and more vapour condenses into liquid water. This reduces the rate at which the rising air cools to the **saturated adiabatic lapse rate** of 1.5°C/1,000 ft.

In this example, at this new rate of cooling the parcel of air will have cooled to the same temperature as the surrounding environment (8°C) at a height of 4,000 ft agl, and so will cease rising. A cumulus cloud, base 2,000 ft and tops 4,000 ft, has been formed.

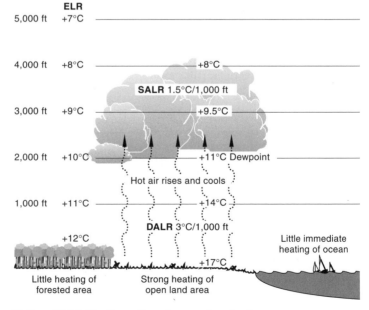

■ *Figure 21-14* **The formation of a cumulus cloud**

Cloud Formed by Orographic Uplift

Air flowing over mountains rises and cools adiabatically. If it cools to below its **dewpoint temperature**, then the water vapour will condense and cloud will form.

Descending on the other side of the mountains, however, the airflow will warm adiabatically and, once its temperature exceeds the dewpoint for that parcel of air, the water vapour will no longer condense. The liquid water drops will now start to vaporise, and the cloud will cease to exist below this level.

A cloud that forms as a 'cap' over the top of a mountain is known as **lenticular** cloud. It will remain more or less stationary whilst the air flows through it.

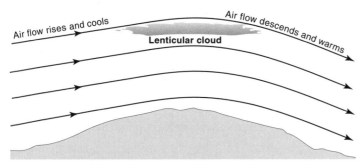

■ *Figure 21-15* **Lenticular cloud as a cap over a mountain**

Sometimes, when an air stream flows over a mountain range and there is a stable layer of air above, **standing waves** occur. This is a wavy pattern as the airflow settles back into a more steady flow and, if the air is moist, lenticular clouds may form in the crest of the lee waves, and a **rotor** or **roll cloud** may form at a low level.

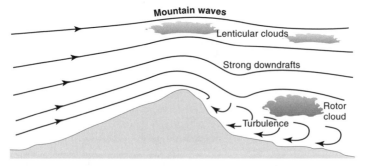

■ *Figure 21-16* **Mountain waves**

The level at which the cloud base forms depends on the moisture content of the parcel of air and its dewpoint temperature. The cloud base may be below the mountain tops or well above them, depending on the situation. Once having started to form, the cloud may sit low over the mountain as stratiform cloud (if the air is stable), or (if the air is unstable) the cloud will be cumuliform and may rise to high levels.

For interesting reading regarding *'Flight Over and in the Vicinity of High Ground'*, refer to Aeronautical Information Circular (AIC) 6/2003 (Pink 48).

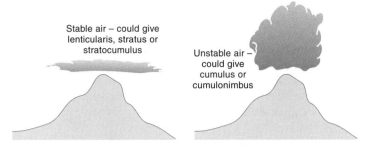

Stable air – could give lenticularis, stratus or stratocumulus

Unstable air – could give cumulus or cumulonimbus

■ *Figure 21-17* **Orographic uplift can lead to cloud formation**

THE FÖHN WIND EFFECT. If the air rising up a mountain range is moist enough to have a high dewpoint temperature and is cooled down to it before reaching the top of the mountain, then cloud will form on the windward side. If any precipitation occurs, moisture will be removed from the airflow and, as it descends on the lee side of the mountain, it will therefore be drier. The dewpoint temperature will be less and so the cloud base will be higher on the lee side of the mountain.

As the dry air beneath the cloud descends, it will warm at the dry adiabatic lapse rate of $3°C/1,000$ ft, which is at a greater rate than the rising air cooled inside the cloud (saturated adiabatic lapse rate: $1.5°C/1,000$ ft). The result is a warmer and drier wind on the lee side of the mountains.

This very noticeable effect is seen in many parts of the world, for example the **föhn** (or *foehn,* pronounced "fern") wind in Switzerland and Southern Germany, from which this effect gets its name.

More locally for the UK, the föhn effect is commonly experienced in North Wales, due to the interaction of south-westerly winds with the mountains of Snowdonia.

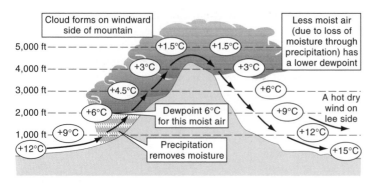

■ *Figure 21-18* **The föhn wind effect**

Cloud Formed by Turbulence and Mixing

As air flows over the surface of the earth, frictional effects cause variations in local wind strength and direction. Eddies are set up which cause the lower levels of air to mix – the stronger the wind and the rougher the earth's surface, the larger the eddies and the stronger the mixing.

The air in the rising currents will cool and, if the turbulence extends to a sufficient height, it may cool to the dewpoint temperature, water vapour will condense to form liquid water droplets and cloud will form.

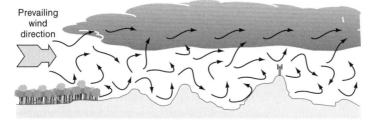

■ *Figure 21-19* **Formation of turbulence cloud**

The descending air currents in the turbulent layer will warm and, if the air's dewpoint temperature is exceeded, the liquid water droplets that make up the cloud will return to the water vapour state. The air will dry out and cloud will not exist below this level.

With turbulent mixing, stratiform cloud may form over quite a large area, possibly with an undulating base. It may be continuous stratus or broken stratocumulus.

Cloud formed by Widespread Ascent

When two large masses of air of differing temperatures meet, the warmer and less dense air will flow over (or be undercut by) the cooler air. As the warmer air mass is forced aloft it will cool and, if the dewpoint temperature is reached, cloud will form.

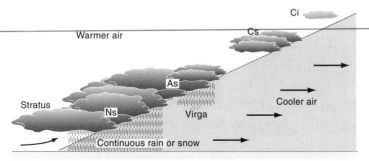

■ *Figure 21-20* **Cloud formation due to widespread ascent**

The boundary layer between two air masses is called a **front**. Weather associated with frontal activity is covered in Chapter 23.

Precipitation Associated with Cloud

Precipitation refers to falling water that finally reaches the ground, including:

- **rain** consisting of liquid water drops;
- **drizzle** consisting of fine water droplets;
- **snow** consisting of branched and star-shaped ice crystals;
- **hail** consisting of small balls of ice;
- **freezing rain or drizzle** which freezes on contact with a cold surface (which may be the ground or an aircraft in flight).

Continuous rain or snow is often associated with nimbostratus and altostratus clouds and intermittent rain or snow with altostratus or stratocumulus. Rain or snow showers are associated with cumuliform clouds such as cumulonimbus, cumulus and altocumulus, extremely heavy showers and/or hail coming from cumulonimbus.

Fine drizzle or snow is associated with stratus and stratocumulus.

It is possible to use precipitation as a means of identifying the cloud type – showers generally falling from cumuliform clouds and non-showery precipitation from stratiform clouds, mainly altostratus and nimbostratus.

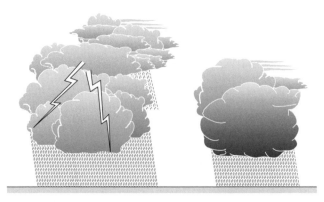

■ *Figure 21-21* **Showers fall from cumuliform clouds**

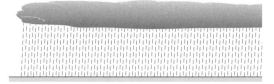

■ *Figure 21-22* **Non-showery precipitation from stratiform clouds**

Rain that falls from the base of clouds but evaporates before reaching the ground (hence is not really precipitation) is called **virga.** It can, of course, affect aircraft which fly through it.

■ *Figure 21-23* **Virga**

Cloud Description in Forecasts and Reports

Aerodromes

The amount and height of clouds in **aerodrome** forecasts (known as TAFs) and aerodrome meteorological reports (known as METARs) follows an international convention, where:

CLOUD AMOUNT ABBREVIATIONS (EXCEPT CB)	
SKC	Sky clear
FEW	Few (1–2 oktas, or eighths of sky covered)
SCT	Scattered (3–4 oktas)
BKN	Broken (5–7 oktas)
OVC	Overcast (8 oktas), sky is completely covered with the cloud layer

The base of the layers or masses of cloud is given in feet above aerodrome level (aal), in ascending order of height. For example, **SCT005 SCT012 BKN050** indicates 3–4 oktas of cloud, base 500 ft aal; with another layer of the same amount at 1,200 ft aal; plus 5–7 oktas at 5,000 ft aal.

The cloud type is not given in TAFs and METARs; however, significant convective clouds (CB: cumulonimbus and TCU: towering cumulus) are appended after the cloud statement, e.g. **SCT008CB.**

NOTE The military, and civil authorities in some countries (e.g. Australia), show cloud in TAFs and METARs using a slightly different system. For example, **3ST004 6SC035** indicates 3 oktas of stratus, base 400 ft aal; and 6 oktas of stratocumulus at base 3,500 ft aal.

Area Forecasts

UK teletype Area Forecasts show the cloud in oktas, along with its type, and height of the base and tops; e.g. **5/8AC 10000FT/18000** indicates 5 oktas of altocumulus cloud at base 10,000 ft, tops 18,000 ft.

NOTE Area Forecasts, TAFs and METARs are discussed in Chapter 26.

Now complete **Exercises 21 – Cloud.**

Thunderstorms

Thunderstorms generate spectacular weather which may be accompanied by thunder, lightning, heavy rain showers and sometimes hail, squalls and tornadoes. Thunderstorms are associated with cumulonimbus clouds and there may be several thunderstorm 'cells' within the one cloud. They constitute a severe hazard to the aviator, especially in light aircraft.

Associated with thunderstorms is lightning, which is a discharge of static electricity that has been built up in the cloud. The air along the path that the lightning follows experiences intense heating. This causes it to expand violently, and it is this expansion which produces the familiar clap of thunder.

The Three Necessary Conditions

For a thunderstorm to develop there must be deep instability, high moisture content and a trigger action.

Three conditions are necessary for a thunderstorm to develop:

☐ **deep instability** in the atmosphere, so that once the air starts to rise it will continue rising (for example, a steep lapse rate with warm air in the lower levels and cold air in the upper levels);

☐ a **high moisture** content;

☐ a **trigger action** (or catalyst) to start the air rising, from:
  – a front forcing the air aloft;
  – a mountain forcing the air aloft;
  – strong heating of the air in contact with the earth's surface; or
  – heating of the lower layers of a polar air mass as it moves to lower latitudes (i.e. towards the equator).

The Life Cycle of a Thunderstorm

1. The Cumulus Stage

As the moist air rises, it is cooled until its dewpoint temperature is reached. Then the water vapour starts to condense as liquid droplets and cloud forms. Latent heat is given off in the condensation process and so the rising air cools at a lesser rate. At this early cumulus stage in the formation of a thunderstorm, there are strong, warm updrafts over a diameter of one or two miles, with no significant downdrafts.

Air is drawn horizontally into the cell at all levels and causes the updraft to become stronger with height. The temperature inside the cloud is higher than the outside environment and the cloud continues to build to greater and greater heights. This often occurs at such a rate that an aeroplane cannot outclimb the growing cloud.

The strong, warm updrafts carry the water droplets higher and higher, to levels often well above the freezing level, where they may freeze or continue to exist as liquid water in a supercooled state. The liquid droplets will coalesce to form larger and larger drops. The cumulus stage typically lasts 10 to 20 minutes.

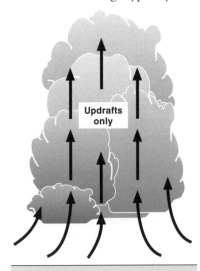

■ *Figure 22-1* **The cumulus stage in the development of a thunderstorm**

2. The Mature Stage

The water drops eventually become too large and heavy to be supported by the updrafts (even though the updrafts may be in excess of 5,000 ft/min) and so start to fall. As they fall in great numbers inside the cloud, they drag air along with them causing **downdrafts**. Often the first lightning flashes and the first rain from the cloud base will occur at this stage.

The descending air warms adiabatically, but the very cold drops of water slow down the rate at which this occurs, resulting in very cool downdrafts in contrast to the warm updrafts. Heavy rain or hail may fall from the base of the cloud, generally being heaviest for the first 5 minutes.

The top of the cloud in this mature stage may reach as far as the tropopause, being some 20,000 ft in temperate latitudes and 50,000 ft in the tropics. The cloud may have the typical shape of a cumulonimbus, with the top spreading out in an *anvil* shape in the direction of the upper winds. The passage of an aeroplane through the windshear of these violent updrafts and downdrafts (which are very close to each other) can result in structural failure.

> **Beware of strong windshear near mature Cb clouds.**

The rapidly changing direction from which the airflow strikes the wings could also result in a stall, so intentionally flying into a mature cumulonimbus cloud would be extremely foolhardy.

As the cold downdrafts flow out of the base of the cloud (at a great rate) they change direction and begin to flow horizontally as the ground is approached. Strong windshear will occur. This has caused the demise of many aeroplanes. The outflowing cold air will undercut the inflowing warmer air and, like a mini cold front, a gusty wind and a sudden drop in temperature may precede the actual storm.

A **roll cloud** may also develop at the base of the main cloud where the cold downdrafts and warm updrafts pass.

The mature stage typically lasts from 20 to 40 minutes.

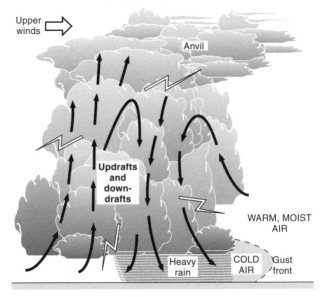

■ *Figure 22-2* **The mature stage**

3. The Dissipating Stage
The cold downdrafts gradually cause the warm updrafts to weaken and so reduce the supply of warm, moist air to the upper levels of the cloud. The cold downdrafts continue (since they are colder than the environment surrounding the cloud) and spread out over the whole cloud, which starts to collapse from above.

Eventually the temperature inside the cloud warms to reach that of the environment and what was once a towering cumulonimbus may collapse into a stratiform cloud.

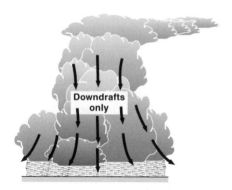

■ Figure 22-3 **The dissipating stage**

The Danger of Thunderstorms

Thunderstorms are hazardous to aviation. The dangers to aviation from a thunderstorm do not exist just inside or under the cloud, but for quite some distance around it. Avoid thunderstorms by at least 10 nm and, in severe situations and higher altitudes, by 20 nm or more. Most jet transports and advanced aeroplanes are equipped with weather radar to enable their pilots to do this. Visual pilots without weather radar have to use their eyes and common sense.

Avoid thunderstorms by at least 10 nautical miles.

Some obvious dangers to aeroplanes from thunderstorms include:
- **severe windshear** (causing flight path deviations and handling problems, loss of airspeed and possibly structural damage);
- **severe turbulence** (causing loss of control and structural damage);
- **severe icing** (possibly the very dangerous clear ice formed from large supercooled water drops striking a sub-zero surface);
- **damage from hail** (to the airframe and cockpit windows);
- **reduced visibility;**
- **damage from lightning strikes,** including electrical damage;
- **interference to radio communications** and radio navigation instruments.

For further reading see AIC 72/2001 (Pink 22).

Now complete **Exercises 22 – Thunderstorms.**

Air Masses and Frontal Weather

Air Masses

An air mass is a large parcel of air with fairly consistent properties (such as temperature and moisture content) throughout. It is usual to classify an air mass according to:

▢ its **origin**;
▢ its **path** over the earth's surface; and
▢ whether the air is **diverging** or **converging**.

The Origin of an Air Mass

Maritime air flowing over an ocean will absorb moisture and tend to become saturated in its lower levels. **Continental air** flowing over a land mass will remain reasonably dry since little water is available for evaporation.

The Path of an Air Mass

Polar air flowing towards the lower latitudes will be warmed from below and so become unstable. Conversely, **tropical air** flowing to higher latitudes will be cooled from below and so become more stable.

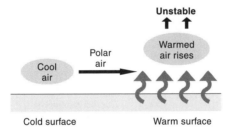

■ *Figure 23-1 **Polar air warms and becomes unstable***

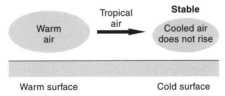

■ *Figure 23-2 **Tropical air cools and becomes stable***

Divergence or Convergence

An air mass influenced by the divergence of air flowing out of a *high*-pressure system at the earth's surface will slowly sink (known as *subsidence*) and become warmer, drier and more stable. An air mass influenced by convergence as air flows into a *low*-pressure system at the surface will be forced to rise slowly, becoming cooler, moister and less stable.

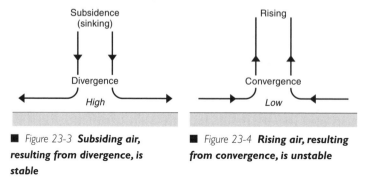

■ *Figure 23-3* **Subsiding air, resulting from divergence, is stable**

■ *Figure 23-4* **Rising air, resulting from convergence, is unstable**

The sources of most air masses that affect the United Kingdom are shown below, classified by temperature and moisture level.

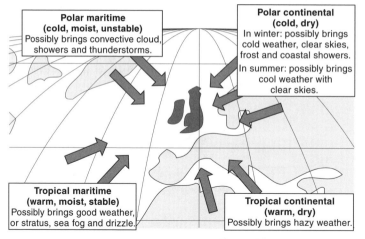

Polar maritime (cold, moist, unstable) Possibly brings convective cloud, showers and thunderstorms.

Polar continental (cold, dry) In winter: possibly brings cold weather, clear skies, frost and coastal showers. In summer: possibly brings cool weather with clear skies.

Tropical maritime (warm, moist, stable) Possibly brings good weather, or stratus, sea fog and drizzle.

Tropical continental (warm, dry) Possibly brings hazy weather.

■ *Figure 23-5* **Air masses that affect the United Kingdom**

Frontal Weather

Air masses have different characteristics, depending on their origin and the type of surface over which they have been passing. Because of these differences there is usually a distinct division between adjacent air masses. These divisions are known as **fronts,** and there are two basic types – cold fronts and warm fronts. *Frontal activity* describes the interaction between the air masses, as one mass replaces the other.

The Warm Front

If two air masses meet so that the warmer air replaces the cooler air at the surface, a **warm front** is said to exist. The boundary at the earth's surface between the two air masses is represented on a weather chart as a line with semicircles pointed in the direction of movement.

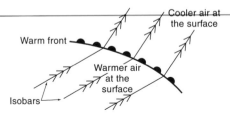

■ *Figure 23-6* **Depiction of a warm front on a weather chart**

The slope formed in a warm front as the warm air slides up over the cold air is fairly shallow and so the cloud that forms in the (usually quite stable) rising warm air is likely to be stratiform. In a warm front the frontal air at altitude is actually well ahead of the line as depicted on the weather chart. The cirrus could be some 600 nm ahead of the surface front, and rain could be falling up to approximately 200 nm ahead of it. The slope of the warm front is typically 1 in 150, much flatter than a cold front, and has been exaggerated in the diagram.

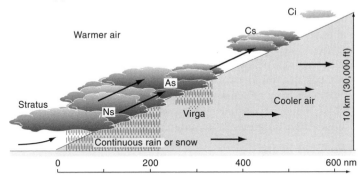

■ *Figure 23-7* **Cross-section of a warm front**

Observation from the Ground

As a warm front gradually passes, an observer on the ground may first see high cirrus cloud, which will slowly be followed by a lowering base of cirrostratus, altostratus and nimbostratus. Figure 21-3 shows an example.

Rain may be falling from the altostratus and possibly evaporating before it reaches the ground – this is called *virga*. Rain may fall continuously from the nimbostratus until the warm front passes and may, due to its evaporation, cause fog. Also, the visibility may be quite poor.

The atmospheric pressure usually will fall continuously as the warm front approaches and, as it passes, either stop falling or fall at a lower rate. The air temperature will rise as the warm air moves in over the surface. The warm air will hold more moisture than the cold air, and the dewpoint temperature in the warmer air will be higher.

In the northern hemisphere, the wind direction will veer as the warm front passes (and back in the southern hemisphere). Behind the warm front, and after it passes, there is likely to be stratus. Visibility may still be poor. Weather associated with a warm front may extend over several hundred miles.

The general characteristics of a warm front are:
- lowering stratiform cloud;
- increasing rain, with the possibility of poor visibility and fog;
- a falling pressure that slows down or stops;
- a wind that veers; and
- a temperature that rises.

Observation from the Air

What a pilot sees, and in which order, will depend on the direction of flight. You may see a gradually lowering cloud base if in the cold sector underneath the warm air and flying towards the warm front, and steady rain may be falling.

If the aeroplane is at sub-zero temperatures, the rain may freeze and form ice on the wings, thereby decreasing their aerodynamic qualities. The cloud may be as low as ground level (i.e. hill fog) and sometimes the lower layers of stratiform cloud conceal cumulonimbus and thunderstorm activity.

Visibility may be quite poor.

There will be a wind change either side of the front and a change of heading may be required to maintain track.

The Cold Front

If a cooler air mass undercuts a mass of warm air and displaces it at the surface, a cold front is said to occur. The slope between the two air masses in a cold front is generally quite steep (typically 1 in 50) and the frontal weather may occupy a band of only 30 to 50 nautical miles.

The boundary between the two air masses at the surface is shown on weather charts as a line with barbs pointing in the direction of travel of the front. The cold front moves quite rapidly, with the cooler frontal air at altitude lagging behind that at the surface.

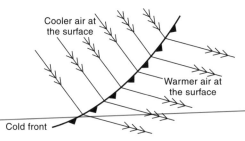

■ *Figure 23-8* **Depiction of a cold front on a weather chart**

The air that is forced to rise with the passage of a cold front is unstable and so the cloud that is formed is cumuliform in nature, e.g. cumulus and cumulonimbus. Severe weather hazardous to aviation, such as thunderstorm activity, squall lines, severe turbulence and windshear, may accompany the passage of a cold front.

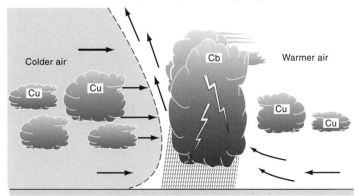

■ *Figure 23-9* **Cross-section of a cold front**

Observation from the Ground

The atmospheric pressure will fall as a cold front approaches and the change in weather with its passage may be quite pronounced. There may be cumulus and possibly cumulonimbus cloud with heavy rain showers, thunderstorm activity and squalls, with a sudden drop in temperature and change in wind direction as the front passes (veering in the northern hemisphere, backing in the southern hemisphere).

The cooler air mass will contain less moisture than the warm air, and so the dewpoint temperature after the cold front has passed will be lower. Once the cold front has passed, the pressure may rise rapidly.

The general characteristics of a cold front are:
- [ ] cumuliform cloud – cumulus, cumulonimbus;
- [ ] a sudden drop in temperature, and a lower dewpoint temperature;
- [ ] a veering of the wind direction; and
- [ ] a falling pressure that rises once the front is past.

Observation from the Air

Flying through a cold front may require diversions to avoid weather. There may be thunderstorm activity, violent winds (horizontal and vertical) from cumulonimbus clouds, squall lines, windshear, heavy showers of rain or hail, and severe turbulence. Icing could be a problem. Visibility away from the showers and the cloud may be quite good, but it is still prudent to consider avoiding the strong weather activity that accompanies many cold fronts.

The Occluded Front

Because cold fronts usually travel much faster than warm fronts it often happens that a cold front overtakes a warm front, creating an **occlusion** or **occluded front**. This may happen in the final stages of a frontal depression (which is discussed shortly). Three air masses are involved and their vertical passage, one to the other, will depend on their relative temperatures. The occluded front is depicted by a line with alternating barbs and semicircles pointing in the direction of motion of the front.

The two types of occluded fronts are illustrated in Figure 23-11. A cold front occlusion will occur when the original cold front remains at the surface. A warm front occlusion will occur when the original warm front remains at the surface. Which of the original fronts remains at the surface of the occlusion will depend on the relative temperatures of the three air masses involved.

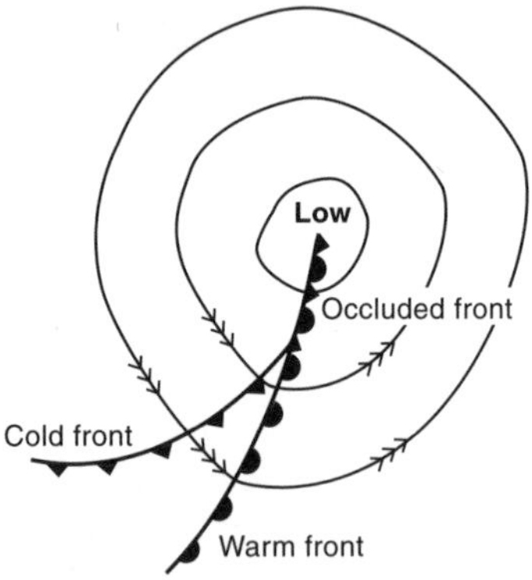

In general the cold front occlusion is a summer phenomenon and the warm front occlusion is more likely to occur in winter.

■ Figure 23-10 **Depiction of an occluded front on a weather map**

The cloud that is associated with an occluded front will depend on what cloud is associated with the individual cold and warm fronts. It is not unusual to have cumuliform cloud (Cu, Cb) from the cold front as well as stratiform cloud from the warm front. Sometimes the stratiform cloud can conceal thunderstorm activity.

Severe weather can occur in the early stages of an occlusion as unstable air is forced upwards, but this period is often short.

Flight through an occluded front may involve encountering intense weather, as both a cold front and a warm front are involved with a warm air mass being squeezed up between them. The wind direction will be different either side of the front.

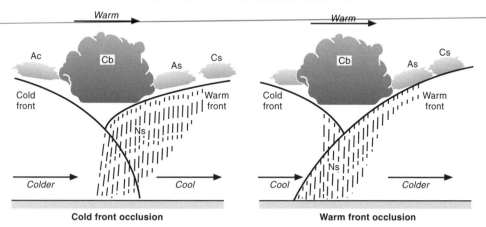

Cold front occlusion   **Warm front occlusion**

■ *Figure 23-11* **Cross-sections of occluded fronts**

ATLANTIC WEATHER – February 16

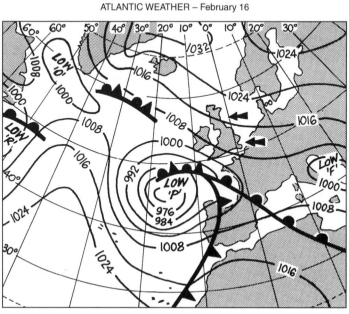

Lows 'P' and 'F' will move east and slowly fill as low 'R' tracks northeast and deepens. Low 'O' will lose its identity.

■ *Figure 23-12* **A weather map showing cold, warm and occluded fronts**

Depressions – Areas of Low Pressure

A **depression** or **low** is a region of low pressure at the surface, the pressure gradually rising as you move away from its centre. A *low* is depicted on a weather chart by a series of concentric isobars joining places of equal sea level pressure, with the lowest pressure in the centre.

In the northern hemisphere, winds circulate anticlockwise around a *low.* Flying towards a *low,* an aeroplane will experience right (starboard) drift.

Depressions generally are more intense than *highs,* being spread over a smaller area and with a stronger pressure gradient (change of pressure with distance). The more intense the depression, the 'deeper' it is said to be. *Lows* move faster across the face of the earth than *highs* and do not last as long.

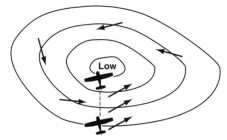

■ *Figure 23-13* **A depression or low pressure system**

Because the pressure at the surface in the centre of a depression is lower than in the surrounding areas, there will be an inflow of air, known as **convergence.** The air above the depression will rise and flow outwards.

The three-dimensional pattern of airflow near a depression is:
☐ convergence (inflow) in the lower layers;
☐ rising air above; and
☐ divergence (outflow) in the upper layers.

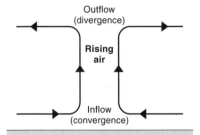

■ *Figure 23-14* **The three-dimensional flow of air near a low**

The depression at the surface may in fact be caused by the divergence aloft removing air faster than it can be replaced by convergence at the surface.

Weather Associated with a Depression

In a depression, the rising air will be cooling and so cloud will tend to form. Instability in the rising air may lead to quite large vertical development of cumuliform cloud accompanied by rain showers. Visibility may be good (except in the showers), since the vertical motion will tend to carry away all the particles suspended in the air.

Troughs of Low Pressure

A V-shaped extension of isobars from a region of low pressure is called a **trough**. Air will flow into it (i.e. convergence will occur) and rise. If the air is unstable, weather similar to that in a depression or a cold front will occur, e.g. cumuliform cloud, possibly with cumulonimbus and thunderstorm activity.

The trough may in fact be associated with a front. Less prominent troughs, possibly more U-shaped than V-shaped, will generally have less severe weather.

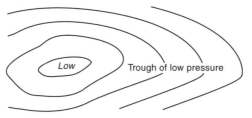

■ Figure 23-15 **A trough**

The Wave or Frontal Depression

The boundary between two air masses moving (relative to one another) side by side is often distorted by the warmer air bulging into the cold air mass, with the bulge moving along like a wave. This is known as a **frontal wave**. The leading edge of the bulge of warm air is a warm front and its rear edge is a cold front.

The pressure near the tip of the wave falls sharply and so a depression forms, along with a warm front, a cold front, and possibly an occlusion. It is usual for the cold front to move faster across the surface than the warm front.

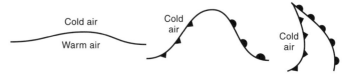

■ Figure 23-16 **The frontal depression**

The Cyclone or Tropical Revolving Storm

The tropical revolving storm can be violent and destructive. Fortunately they do not occur in the UK, but rather over warm tropical oceans at about 10–20° latitude during certain periods of the year.

Occasionally, weak troughs in these areas develop into intense depressions. Air converges in the lower levels and flows into the depression and then rises, the warm, moist air forming large cumulus and cumulonimbus clouds. The very deep depression may be only quite small (200–300 nm in diameter) compared to the typical depression in temperate latitudes, but its central pressure can be extremely low.

Winds can exceed 100 kt, with heavy showers and thunderstorm activity becoming increasingly frequent as the centre of the storm approaches.

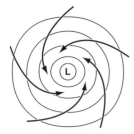

■ *Figure 23-17* **A tropical revolving**
storm or cyclone

The **eye** of a tropical revolving storm is often only some 10 nm in diameter, with light winds and broken cloud. It is occupied by very warm subsiding air, one reason for the extremely low pressure. Once the eye has passed, a very strong wind from the opposite direction will occur.

In the northern hemisphere, pronounced starboard (right) drift due to a strong wind from the left will mean that the eye of the storm is ahead (and vice versa in the southern hemisphere).

Tropical revolving storms are known as *cyclones* in the Indian and Pacific Ocean areas, as *hurricanes* in the Caribbean, and *typhoons* in the China Sea area.

These intense weather systems are best avoided by all aircraft.

Anticyclones – Areas of High Pressure

An **anticyclone,** or **high,** is an area of high pressure at the surface surrounded by roughly concentric isobars. *Highs* are generally greater in extent than *lows*, but with a weaker pressure gradient and slower moving, although they are more persistent and last longer.

In the northern hemisphere, the wind circulates clockwise around the centre of a *high*. Flying towards a *high* an aircraft will experience left (port) drift.

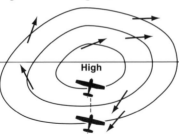

■ *Figure 23-18* **The anticyclone or high**

The **three-dimensional flow** of air associated with an anti-cyclone is:

☐ an outflow of air from the high-pressure area in the lower layers (divergence);

☐ the slow subsidence of air over a wide area from above; and

☐ an inflow of air in the upper layers (convergence).

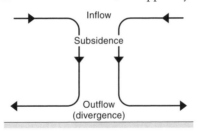

■ *Figure 23-19* **The three-dimensional flow of air near a high**

The high-pressure area at the surface originates when the convergence in the upper layers adds air faster than the divergence in the lower layers removes it.

Weather Associated with a High

The subsiding air in a high-pressure system will be warming as it descends and so cloud will tend to disperse as the dewpoint temperature is exceeded and the relative humidity decreases. Subsiding air is very stable.

It is possible that the subsiding air may warm sufficiently to create an inversion, with the upper air warming to a temperature higher than that of the lower air, and possibly causing stratiform cloud to form (stratocumulus, stratus) and/or trapping smoke, haze and dust beneath it. This can happen in the UK winter, leading to rather gloomy days with poor flight visibility. In summer, heating by the sun may disperse the cloud, leading to a fine but hazy day.

If the sky remains clear at night, greater cooling of the earth's surface by radiation heat loss may lead to the formation of fog. If the high pressure is situated entirely over land, the weather may be dry and cloudless, but with any air flowing in from the sea, extensive stratiform cloud in the lower levels can occur.

A Ridge of High Pressure

Isobars which extend out from a *high* in a U-shape indicate a ridge of high pressure (like a ridge extending from a mountain). Weather conditions associated with a ridge are, in general, similar to the weather found with anticyclones.

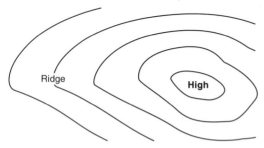

■ *Figure 23-20* **A ridge**

A Col

The area of almost constant pressure (and therefore indicated by a few very widely-spaced isobars) that exists between two *highs* and two *lows* is called a **col**. It is like a 'saddle' on a mountain ridge.

Light winds are often associated with cols, with fog a possibility in winter and high temperatures in summer possibly leading to showers or thunderstorms.

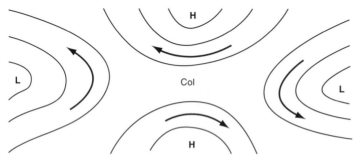

■ *Figure 23-21* **A col**

Now complete **Exercises 23 – Air Masses and Frontal Weather.**

Icing

Dangers of Icing

Ice accretion on an aeroplane or within the engine induction system can significantly reduce flight safety by causing:

☐ **Adverse aerodynamic effects** – ice accretion on the *airframe* can modify the airflow pattern around aerofoils (wings, propeller blades, etc.), leading to a serious loss of lift and an increase in drag.

☐ **A loss of engine power**, or even complete stoppage, if ice blocks the air intake (in sub-zero temperatures) or carburettor ice forms (in moist air up to +25°C).

☐ **A weight increase** and a **change in the CG position** of the aeroplane, as well as unbalancing of the various control surfaces and the propeller, perhaps causing severe vibration and/or control difficulties.

☐ **Blockage of the pitot tube and/or static vent**, producing errors in the pressure instruments (airspeed indicator, altimeter, vertical speed indicator).

☐ **Degradation in radio communications** (if ice forms on the aerials).

The Formation of Ice

If the temperature is less than 0°C, which is the freezing point of water, then ice may form, either:

☐ directly from water vapour (sublimation, causing hoar frost); or
☐ from water droplets freezing, causing rime ice and/or clear ice.

Supercooled water drops

Liquid water drops can exist in the atmosphere at temperatures well below the normal freezing point of water (0°C), possibly at −20°C or even lower. This is known as being *supercooled,* and such drops will freeze on contact with a surface – the skin of an aeroplane, or the propeller blades, for example.

Clear Ice

Once the freezing process actually begins, a large water drop with a temperature between 0°C and −20°C will not freeze instantaneously. The freezing process could be triggered by the drop striking a cold aircraft surface, where it will start to freeze. Because latent heat is released in this process, the rate of freezing will decrease, and the remaining liquid water will spread back and coalesce with water from other partially frozen water drops, before freezing on the cold airframe or propeller surfaces. The result is a sheet of solid, clear, glazed ice with very little air enclosed.

■ *Figure 24-1* **Clear ice formed from large, supercooled water drops**

The surface of clear ice is smooth, usually with undulations and lumps. Clear ice can alter the aerodynamic shape of aerofoils quite dramatically and reduce or destroy their effectiveness. Along with the increased weight, this creates a hazard to safety. Clear ice is very tenacious but, if it does break off, it could be in large chunks capable of doing damage.

> *Clear ice is a major hazard to flight safety.*

Rime Ice

Rime ice occurs when small, supercooled liquid water droplets freeze on contact with a surface whose temperature is sub-zero. Because the drops are small, the amount of water remaining after the initial freezing is insufficient to coalesce into a continuous sheet before freezing. The result is a mixture of tiny ice particles and trapped air, giving a rough, opaque, crystalline deposit that is fairly brittle.

> *Rime ice is the most common form of icing.*

Rime Ice often forms on leading edges and can affect the aerodynamic qualities of an aerofoil or the airflow into the engine intake. It does cause a significant increase in weight.

Cloudy or Mixed Ice

It is common for the drops of water in falling rain to be of many sizes and often, if ice forms, it will be a mixture of clear ice (from large drops) and rime ice (from small drops), resulting in cloudy or mixed ice.

Hoar Frost

Hoar frost occurs when moist air comes in contact with a sub-zero surface, the water vapour, rather than condensing to form liquid water, sublimating directly to ice in the form of hoar frost. This is a white crystalline coating that can usually be brushed off.

Hoar frost will form in clear air when the aeroplane is parked in sub-zero temperatures or when the aeroplane flies from sub-zero temperatures into warmer moist air – for example, on descent, or when climbing in an inversion. Although hoar frost is not as dangerous as clear ice, it can obscure vision through a cockpit window, and possibly affect the lifting characteristics of the wings.

> *Frost remaining on the wings is dangerous, especially during take-off.*

Structural Icing and Cloud Type

Ice adhering to the airframe is a very important consideration for instrument-rated pilots who may be flying in cloud; it is also an important consideration for visual pilots who may be flying in rain or drizzle which freezes on a cold aeroplane. Carburettor ice, which is discussed shortly, can of course occur without the presence of cloud or precipitation.

Cumulus-type cloud nearly always consists predominantly of liquid water droplets down to about −20°C, below which either liquid drops or ice crystals may predominate. Newly formed parts of the cloud will tend to contain more liquid drops than mature parts. The risk of airframe icing is severe in cumuliform cloud in the range 0 to −20°C, moderate to severe in the range −20° to −40°C, with the chance of airframe icing below −40°C being only small. Since there is a lot of vertical motion in convective clouds, the composition of the clouds may vary considerably at the one level, and the risk of icing may exist throughout a wide altitude band in (and under) the cloud. If significant icing does occur, it may be necessary to descend into warmer air.

Stratiform cloud usually consists entirely or predominantly of liquid water drops down to about −15°C, with the risk of airframe icing. If significant icing is a possibility, it may be advisable to fly at a lower level where the temperature is above 0°C, or at a higher level where the temperature is less than −15°C. In certain conditions, such as stratiform cloud associated with an active front or with orographic uplift, the risk of icing is increased at temperatures lower than usual; continuous upward motion of air generally means a greater retention of liquid water in the cloud.

Raindrops and drizzle from any sort of cloud will freeze if they meet an aeroplane whose surface is below 0°C, with a severe risk of clear ice forming, the bigger the water droplets are. You need to be cautious when flying in rain at freezing temperatures. This could occur, for instance, with an aeroplane flying in the cool sector underlying the warmer air of a warm front from which rain could be falling.

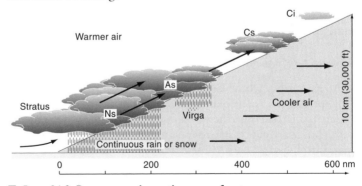

■ Figure 24-2 **Danger area beneath a warm front**

Cirrus clouds are usually composed of ice crystals, and the risk of airframe icing is therefore only slight.

Carburettor Icing

Ice can form in the carburettor and induction system in **moist air** with outside air temperatures as high as 30°C. It will disturb or prevent the flow of air and fuel into the engine, causing it to lose power, run roughly and perhaps even stop.

> When the air is moist, carburettor ice can form in temperatures as high as 30°C!

Cooling occurs when the induction air expands as it passes through the venturi in the carburettor (adiabatic cooling), and occurs also as the fuel vaporises (absorbing the latent heat of vaporisation). This can easily reduce what was initially quite warm air to a temperature well below zero and, if the air is **moist**, ice will form.

Throttle icing is more likely to occur at lower power settings when the partially closed butterfly creates a greater venturi cooling effect, compared with high power settings when the butterfly is more open and the venturi effect is less.

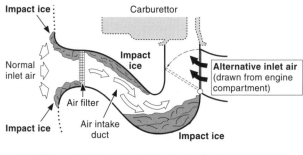

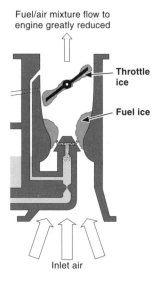

Alternative inlet air

In systems with float-type carburettors this is manual **carburettor heat** air.

In fuel-injection systems this is **alternate air**, usually supplied automatically if normal inlet is blocked by ice.

Note: As carburettor heat/alternate air is not filtered its use on the ground must be kept to a minimum.

■ *Figure 24-3* **Carburettor ice**

All aeroplanes whose engines have carburettors are fitted with a carburettor heat control that can direct hot air from around the engine to be taken into the carburettor, instead of the ambient air. Being hot, it should be able to melt the ice and prevent further ice from forming. The correct method of using the carburettor heat control for your aeroplane will be found in the Pilot's Operating Handbook, and its use is also covered in Volume 1 of this series.

> For more information on carburettor ice, see AIC 145/97 (Pink 161) and Vol. 4 of The Air Pilot's Manual.

Remember that carburettor ice can form on a warm day in moist air!

Pitot-Static System Icing

This can adversely affect the readings of the pressure-operated flight instruments (i.e. the airspeed indicator, the altimeter, and the VSI). Refer to Vol. 4 of *The Air Pilot's Manual,* under *Pressure Instruments.*

Warning!

An ice-laden aeroplane may be completely incapable of flight.

Ice of any type on the airframe or propeller, or in the carburettor and induction system, deserves the pilot's immediate attention and removal. Wings that are contaminated by ice prior to take-off will lengthen the take-off run because of the higher speed needed to fly – a dangerous situation!

An ice-laden aeroplane may even be incapable of flight. Ice or frost on the leading edge and upper forward area of the wings (where the majority of the lift is generated) is especially dangerous.

Most training aeroplanes are not fitted with airframe de-icers (removal) or anti-icers (preventative), so pilots of these aeroplanes should avoid flying in icing conditions (i.e. in rain or moist air at any time the airframe is likely to be at sub-zero temperatures). If a pitot heater is fitted, use it to avoid ice forming over the pitot tube and depriving you of airspeed information.

Now complete **Exercises 24 – Icing.**

Visibility

To non-instrument-rated pilots, flight visibility is one of the most important aspects of weather. You need a natural horizon as a guide to the attitude of the aeroplane in both pitch and bank, as well as having sufficient slant visibility to see the ground for navigational purposes.

Meteorological visibility is defined as the greatest horizontal distance at which a specified object can be seen in daylight conditions. It is a measure of how transparent the atmosphere is to the human eye.

Visibility for a Pilot

The most important visibility to the visual pilot is that from the aeroplane to the ground, i.e. **slant or oblique visibility.** This may be quite different to the horizontal visibility – for example, when a layer of particles is suspended in the air, such as stratus, fog, or smog that is held beneath a temperature inversion.

A common situation is when a ground fog occurs. To an observer on the ground, horizontal visibility might be reduced to just a few hundred metres, yet vertical visibility might be unlimited with a blue sky quite visible above.

To a pilot flying overhead the aerodrome, the runway might be clearly visible and the horizontal visibility ahead unlimited, yet, once he has positioned the aeroplane on final approach, the runway might be impossible to see. This is because his line of sight must now penetrate a much greater thickness of fog.

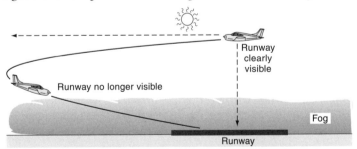

■ *Figure 25-1* **Slant visibility may be severely reduced by fog, smog or stratus**

Particles in the Air

Visibility can be reduced by particles suspended in the air.

On a perfectly clear day visibility can exceed 100 nm; however, this is rarely the case since there are always some particles suspended in the air to prevent all of the light from a distant object reaching the observer.

Particles that restrict visibility include:
- ☐ minute particles of smoke so small that even very light winds can support them:
  - dust or oil causing haze;
  - liquid water or ice producing mist, fog or cloud;
- ☐ larger particles of sand, dust or sea spray which require stronger winds and turbulence for the air to hold them in suspension;
- ☐ precipitation (rain, snow, hail), the worst visibility being associated with very heavy rain or with large numbers of small particles, e.g. thick drizzle or heavy, fine snow.

Rain or snow will reduce the distance that a pilot can see, as well as possibly obscuring the horizon. Poor visibility in the whole area may occur in mist, fog, smog, stratus, drizzle or rain. Unstable air may cause cumuliform clouds to form with poor visibility in the showers falling from them, but good visibility otherwise.

Heavy rain may collect on the windscreen, especially if the aeroplane is flying fast, and cause optical distortions. If freezing occurs on the windscreen, either as ice or frost, vision may also be impaired.

Strong winds can raise dust or sand from the surface and, in some parts of the world, visibility may be reduced to just a few metres in dust and sand storms.

Sea spray often evaporates after being blown into the atmosphere, leaving behind small salt particles that can act as condensation nuclei. The salt particles attract water and can cause condensation at relative humidities as low as 70%, restricting visibility much sooner than would otherwise be the case. Haze produced by sea salt often has a whitish appearance.

Position of the Sun

When flying *down-sun* where the pilot can see the sunlit side of objects, visibility may be much greater than when flying into the sun. As well as reducing the visibility, flying into the sun may cause glare. If landing into the sun is necessary, consideration should be given to altering the time of arrival.

Remember that the onset of darkness is earlier on the ground than at altitude and, even though visibility up high might be good, flying low in the circuit area and approaching to land on a darkening airfield may cause visibility problems.

Inversions and Reduced Visibility

An inversion occurs when the air temperature increases with height (rather than the usual decrease), and this can act as a blanket, stopping vertical convection currents. Air that starts to rise meets warmer air and so will stop rising.

Smog can form beneath an inversion and severely reduce visibility.

Particles suspended in these lower layers will be trapped there and so a rather dirty layer will form, particularly in industrial areas. These small particles may act as **condensation particles** and encourage the formation of fog as well, the combination of smoke and fog being known as **smog**.

Similar effects can be seen in rural areas if there is a lot of pollen, dust or other matter in the air.

Inversions can occur by cooling of the air in contact with the earth's surface overnight, or by subsidence associated with a *high* as descending air warms.

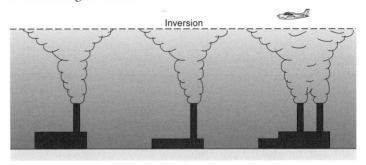

■ *Figure 25-2* **Inversions can lead to reduced visibility**

Mist and Fog

Mist and fog occur when small water droplets are suspended in the air, thereby reducing visibility. The difference between mist and fog is only one of degree – **mist** existing if the visibility exceeds 1 km, **fog** existing if it falls below 1 km.

It is usual for mist to precede fog at a place and to follow fog as it disperses, unless of course an already formed fog is blown in across an area – for example a fog blown in off the sea.

The condensation process that causes mist/fog is usually associated with cooling of the air by an underlying cold surface (causing radiation fog or advection fog) or by the interaction of two air masses (frontal fog). In fog, the relative humidity is 100%; in mist, the relative humidity is slightly less than this.

Radiation Fog

Conditions suitable for the formation of radiation fog are:

- **a cloudless night**, allowing the earth to lose heat by radiation to the atmosphere and thereby cool, also causing the air in contact with it to lose heat;
- **moist air** (i.e. a high relative humidity) that only requires a little cooling to reach the dewpoint temperature; and
- **light winds** (5–7 knots) to mix the lower levels of air with very light turbulence, thereby thickening the mist/fog layer.

These conditions are commonly found with an anticyclone (high-pressure system).

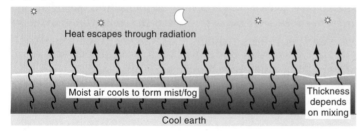

■ *Figure 25-3* **Radiation fog**

Air is a very poor conductor of heat, so that if the wind is absolutely calm, only the very thin layer of air an inch or two thick actually in contact with the surface will lose heat to it. This will cause dew or frost to form on the surface itself, instead of fog forming above it. **Dew** will form at temperatures above zero and **frost** will form at sub-zero temperatures.

If the wind is stronger than about 7 knots, the extra turbulence may cause too much mixing and, instead of fog right down to the ground, a layer of stratus may form above the surface.

■ *Figure 25-4* **Wind strength will affect the formation of dew/frost, mist/fog or stratus cloud**

The temperature of the sea remains fairly constant throughout the year, unlike that of the land which warms and cools quite quickly on a diurnal (daily) basis. Radiation fog is therefore much more likely to form over land than over the sea.

Dispersal of Radiation Fog

As the earth's surface begins to warm up again some time after sunrise, the air in contact with it will also warm, causing the fog to gradually dissipate. In the UK spring it is common for this to occur by early or mid-morning. Possibly the fog may rise to form a low layer of stratus before the sky fully clears.

If the fog that has formed overnight is thick, however, it may act as a blanket, shutting out the sun and impeding the heating of the earth's surface after the sun has risen. As a consequence, the air in which the fog exists will not be warmed from below and the fog may last throughout the day. This is sometimes experienced in the UK during autumn.

Advection Fog

A warm, moist air mass flowing across a significantly colder surface will be cooled from below. If its temperature is reduced to the dewpoint temperature, then fog will form. Since the term *advection* means heat transfer by the horizontal flow of air, fog formed in this manner is known as **advection fog,** and can occur quite suddenly, day or night, if the right conditions exist.

Light to moderate winds will encourage mixing in the lower levels to give a thicker layer of fog, but stronger winds may cause stratus rather than fog. Advection fog can persist in much stronger winds than radiation fog.

A warm, moist maritime airflow over a cold land surface can lead to advection fog over land.

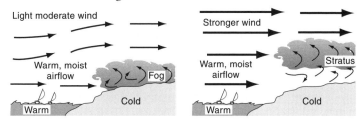

■ *Figure 25-5* **Fog or stratus caused by advection**

Fog at Sea and Along Coastal Areas

Sea fog is advection fog, and it may be caused by:

☐ tropical maritime air moving towards the pole over a colder ocean or meeting a colder air mass; or by

☐ an airflow off a warm land surface moving over a cooler sea, which can occur in the UK summer, affecting aerodromes in coastal areas.

Frontal Fog

This type of fog forms from the interaction of two air masses in one of two ways:

☐ as cloud that extends down to the surface during the passage of the front (forming mainly over hills and consequently called **hill fog**); or

☐ as air becomes saturated by the evaporation from rain that has fallen.

These conditions may develop in the cold air ahead of a warm front (or an occluded front), the pre-frontal fog possibly being very widespread.

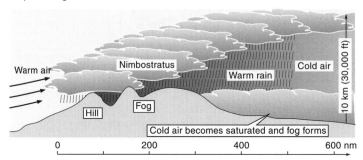

■ Figure 25-6 **Fog associated with a warm front**

Visibility in Forecasts and Reports

Visibility is given to the pilot in kilometres or, if it is very poor, in metres. An Aerodrome Forecast that contains the term **4000 – RADZ** can be interpreted, using a decode table, as "Visibility 4,000 metres in light rain and drizzle". (The decodes are explained in the next chapter.)

If the visibility is 10 km or more (i.e. in excess of 9999 metres), it will be written as '9999'; a visibility of 4 km (or 4,000 metres) will be written as '4000'; a visibility of less than 50 metres will be written as '0000'.

For instrument-rated pilots attempting to operate on runways in very poor visibility, the **runway visual range** (RVR), may be quoted in metres. For instance, **R25/0300** on a Meteorological Observation means "Runway visual range on Runway 25 is 300 metres". It is the visibility along the runway that is important to the pilot, and not the general visibility around the aerodrome, which may be quite different. See UK AIP GEN 3-5.

Now complete **Exercises 25 – Visibility.**

Weather Forecasts and Reports

Since weather conditions vary from place to place and from time to time, it is good airmanship (common sense) to be aware of the weather that you are likely to encounter in the course of a flight, particularly if it is a cross-country flight to another aerodrome.

Checking the weather can be done to a limited extent by making one's own observations; it can be done more adequately, however, by consideration of the information available to pilots from the Meteorological Office. Information will include air temperatures, winds, pressure patterns, the extent and base of any cloud, and the possibility of fog, icing, thunderstorms or other weather phenomena.

Weather information available to pilots falls into two categories:
- **forecasts** of expected weather over an area or region, and at aerodromes; and
- **reports** of actual weather.

Once obtained, this information can be considered carefully pre-flight. It can also be obtained as a flight progresses, to enable the pilot to ensure that the weather conditions are likely to remain suitable for the planned flight, and if not, to enable alternative plans to be made in time.

Dissemination of Weather Information

Weather information is issued by Meteorological Forecast Offices on a routine basis in two formats.
- **Text-based teleprinter AIRMET messages** transmitted to aerodrome briefing offices via the Aeronautical Fixed Telecommunications Network (AFTN), the Facsimile Broadcast Service ARTIFAX, and telex; AIRMET information is also available to individual users via the public telephone network and the dial-up METFAX Service.
- **Graphic weather charts and data** distributed via the Facsimile Broadcast Service to aerodrome briefing offices and via the METFAX Service to individual users.
- **Internet services**, which are available to all PC users who have access to the Internet. These services include the text-based and graphical data already mentioned.

The primary method of obtaining a pre-flight meteorological briefing in the United Kingdom is by **self-briefing**, using:
- facilities, information and documentation routinely available or displayed in aerodrome briefing areas;
- the automated facsimile-based METFAX Service for pilots;

- the Internet services provided by the UK Meteorological Office (www.metoffice.com/aviation) or from other service providers who are authorised to provide the same information via the Internet;
- the public telephone system to obtain AIRMET forecasts and reports; or
- the PC-based service MIST (Meteorological Information Self-briefing Terminal).

Self-briefing does not require prior notification by the pilot. Where necessary, the personal advice of a forecaster or the supply of additional weather information can be obtained from the designated forecast office for the departure aerodrome (listed in AIP GEN 3-5).

All of the Met Office services to pilots are summarised in the publication *Get Met*, which is available free of charge from the Met Office. This booklet is a joint Met Office and CAA publication and may be ordered or downloaded via the Met Office website (www.metoffice.com). *Get Met* complements the information found in AIP GEN 3-5 and is updated annually.

The AIRMET Service

AIRMET is a text-based general-aviation weather briefing service provided by the UK Meteorological Office and made available to both aerodrome briefing offices and to individual users. For individual users, AIRMET is accessed through automated telephone systems that supply:

- low-level Area Forecasts; and
- Aerodrome Reports and Forecasts.

AIRMET Area Forecasts are issued four times daily for three Regions covering the UK and near continent (Scottish, Northern and Southern England). Figure 26-1 shows the AIRMET boundaries and regions.

For pilots using the public telephone network, AIRMET telephone numbers are listed in AIP GEN 3-5, and on the 'AIRMET copy form' (also found in AIP GEN 3-5). Although not essential, using the AIRMET form simplifies the task of copying down the forecasts and enables optimum use of the system. An example of a completed AIRMET copy form is shown in Figures 26-2 and 26-3.

The length of each AIRMET regional Area Forecast varies according to the complexity of the weather situation, and may at times exceed six minutes. The format of the forecast is designed to enable telephone users to determine early in the broadcast if conditions are suitable for VFR operations or not. If conditions are not suitable for the intended flight, the telephone call can be curtailed and only minimal cost incurred.

AIRMET Aerodrome Forecasts (TAFs) and Actual Weather Reports (METARs) are available via telephone, fax or the Internet. Full details may be found in the *Get Met* booklet. UK AIP GEN 3-5 details how to use the AIRMET automated telephone service for obtaining TAFs and METARs. (TAFs and METARs are also available at aerodrome briefing offices.)

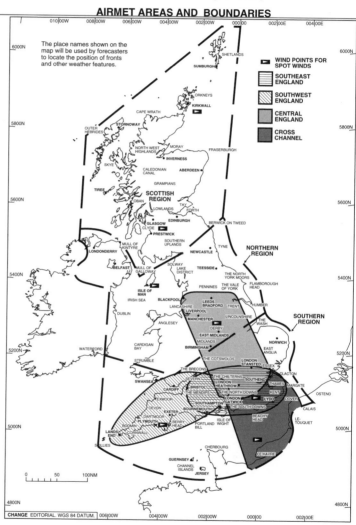

■ Figure 26-1 **AIRMET boundaries and regions**

Perhaps the easiest way to understand the weather service is to use it – by visiting an aerodrome briefing office to obtain the information in text form, and/or by using the public telephone network.

Try phoning the listed telephone numbers and writing down the AIRMET information. You will find the telephone numbers for voice-based forecasts for the Northern, Southern and Scottish Regions in UK AIP GEN 3-5.

Facsimile-Based Automated Weather Services

Together with the Facsimile Broadcast Service ARTIFAX, which distributes weather information to participating aerodrome briefing offices, the dial-up facsimile service called METFAX is available for individual users who have the use of facsimile equipment. METFAX distributes the same information as the fax broadcast service ARTIFAX.

Most modern facsimile systems capable of 'polling' should be able to access the METFAX system. *Get Met* explains the weather services available and how the system works, together with update and validity times, and the telefax numbers and Internet sites to use for accessing each product.

'Get Met' is available free of charge from the Met Office (fax: 0845 3001300; e-mail: metfax@meto.gov.uk).

METFAX supplies an extensive range of products, including:
- Surface Analysis and Forecast charts;
- the UK Low-Level Weather chart (Form 215; see Figure 26-5, page 278);
- Explanatory Notes for Form 215;
- the UK Spot Wind chart (Form 214; see Figure 26-6, page 279);
- 48-hour and 72-hour Surface Outlook charts;
- satellite pictures;
- four European Weather and Spot Wind charts;
- Regional AIRMET Text Forecasts; plus
- Aerodrome Forecasts (TAFs) and Reports (METARs), grouped into various areas of the UK.

Aerodrome Forecasts, Reports and Forecast Charts are now available (free) from the Met Office aviation website (www.metoffice.com/ aviation). For further information contact the Met Office customer centre +44(0) 1344 855680

Examples of two of the dial-up fax products – the UK Low-Level Weather chart and the UK Spot Wind chart – are included later in the chapter under *Area Forecasts*.

Internet Services

Any PC user with access to the Internet may obtain comprehensive briefing services via the Met Office website:
www.metoffice.com/aviation

Most of the commonly used date is available free of charge:
- Forms 214 and 215
- Forms 414 and 415 (European low level spot wind and forecast charts)
- Global TAFs and METARs
- UK AIRMETs
- Radar

- Ballooning forecasts
- Pressure charts
- Climate statistics for British Isles airfields
- Satellite images

Further services are available by subscription:
- 15-minute rainfall radar
- ~~Lightning observations~~
- Animated cloud, precipitation and pressure forecasts
- Animated visibility forecasts
- Animated European synoptic analysis
- Visible satellite pictures
- Infrared satellite pictures
- Three-day planning forecasts for England, Wales and Scotland
- Jetstream chart

Further details may be found in the *Get Met* booklet and on the website.

Forecast Offices

Amplification and consultation on weather information is available direct from the various forecast offices. These are listed in AIP GEN 3-5 and on the AIRMET copy form, with telephone numbers to use.

While the recorded AIRMET messages may be listened to for as long as a pilot wishes, it is requested that phone calls to a Meteorological Office that will involve meteorological personnel be kept reasonably brief.

Types of Weather Information

A **forecast** is a prediction (or prognosis) of what the weather is likely to be – the common aviation forecasts being:
- **Area Forecasts;**
- **Aerodrome Forecasts** (TAFs or TRENDs); and
- **Special Forecasts.**

A **report** is an observation of what the weather actually is (or was) at a specific time. The common aviation weather reports are:
- Aerodrome Weather Reports (**METAR**s on a routine basis and **SPECI**s when special conditions exist);
- Automatic Terminal Information Service (**ATIS**);
- In-flight Weather Reports (obtainable in recorded form on a **VOLMET** VHF frequency, e.g. London VOLMET on 128.600 or 126.600 MHz, or from an ATIS, or by radio communication with an Air Traffic Service unit – although this is the least preferred method since it occupies a communications frequency).

Significant weather that may affect the safety of flight operations may be advised in the form of a **SIGMET**. The criteria for raising a SIGMET include active thunderstorms, tropical revolving storms, a severe line squall, heavy hail, severe turbulence, severe airframe icing, marked mountain waves, widespread dust or a sandstorm.

As a means of improving the meteorological information service to pilots, it is common to attach a forecast **TREND** to an Aerodrome Report (i.e. an observation of actual weather). The Trend Forecast indicates what the weather tendency over the following two hours is expected to be. It is valid only until 2 hours after the time of the observation – a much shorter period than the duration of a normal Aerodrome Forecast (9 hours) – and therefore should be more accurate.

A TREND is commonly referred to as a **landing forecast.**

Meteorological Forecasts

Area Forecasts

An Area Forecast, as the name implies, provides information on expected weather for a certain area over a certain period. It may cover a large area such as the entire British Isles, or a more localised AIRMET region such as Northern England.

AIRMET Forecasts

AIRMET Area Forecasts are text-based. They are suitable for the type of operations that light aircraft commonly undertake in the UK. Also, as it is simply a matter of making an inexpensive phone call or fax request, the service can be obtained easily wherever there is a telephone or suitable facsimile equipment.

AIRMET Area Forecasts cover conditions from the surface (ground or sea level) to 15,000 ft, with winds and air temperatures to 18,000 ft. They are issued four times daily for each region and amended as necessary between those times to account for significant weather changes. Each forecast will include a brief outlook to the end of the subsequent forecast period to give a preview of subsequent weather conditions.

CIVIL AVIATION AUTHORITY

AIRMET COPY FORM
(for forecasts from 31 March 1995 onwards)

Forecast Number [   ] Valid 15/0500 To [   ] 15/1300Z

All temperatures are in degrees Celsius. All heights are AMSL, and all times UTC(Z).

MET SITUATION

SAMPLE ONLY
Not to be used for flight
operations or flight planning

Strong unstable NW flow covers
the region.

Strong wind warning; NW surface
wind will gust to 35 kt in places.

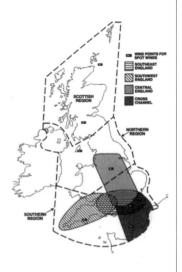

ALTITUDE	WINDS/TEMPERATURES	FORECAST AREA		
1000 Ft	320/30 kt +4°C	1. SOUTHERN REGION	40	☒
3000 Ft	320/35 kt 0°C	2. NORTHERN REGION	41	
		3. SCOTTISH REGION	42	
6000 Ft	320/30 kt −4°C	4. UK SIG WEATHER	43	
10000 Ft		5. UK UPPER WINDS	44	
	FZ level 3000 ft	6. UK UPDATE/OUTLOOK	45	
18000 Ft		7. SOUTHWEST ENGLAND	46	
		8. SOUTHEAST ENGLAND	47	
		9. CROSS CHANNEL	48	
24000 Ft		10. CENTRAL ENGLAND	49	

■ Figure 26-2 **Example of an AIRMET regional forecast (reduced), taken
down using the telephone service, onto the copy form**

Vis gen 30 km, with nil to 3/8 Sc 2500 ft to 6000 ft.

Isol, mainly over sea, and coasts and hills exposed to NW, 10 km in rain showers, with 7/8 Cu 1500 ft to 10,000 ft.

Isol, in the far NE and NW, vis 3000 m in TS and hail, with 7/8 Cb 1000 ft to 20/24,000 ft, and in the NE 5/8 AC layered 10/16,000 ft.

Occ in extreme Sth at first, vis 8 km in rain, with 5 to 7/8 Sc and Ac layered to 1500 ft to 16,000 ft.

WARNING
Cloud covering hills. Mod turb and mod ice in cloud, but isol severe ice and severe turb in Cb in NE and NW.

Mod turb below 6000 ft over land due to strong surface winds.

OUTLOOK   UNTIL   [15/1900Z   ]

More showers developing over land as a trough moves south across region; to lie Cherbourg to Straits of Dover by 1800Z.

TELEPHONE NUMBERS	
All forecasts between 0530 and 2300 on 0891 77 13 plus two digit code for selected forecasts. **TAFs/METARs** ALDERGROVE 018 494 23275 BIRMINGHAM 0121 717 0580 CARDIFF 01222 390 492 LEEDS 01132 457 687 BRACKNELL 01344 856 267 MANCHESTER 0161 429 0927 GLASGOW 0141 221 6116	All forecasts between 2300 and 0530 PLUS Consultation for amplification of all forecasts. BRACKNELL 01344 856 267 MANCHESTER 0161 429 0927 GLASGOW 0141 221 6116

Copies of this form, and of CAA document No 397 AIRMET FORECASTS; USERS NOTES-2nd ISSUE, may be obtained free of charge by writing to: The Technical Secretary, AOPA, 50a Cambridge Street, London SW1V 4QQ, provided a suitable A4 size stamped addressed envelope is enclosed with the request.

■ Figure 26-3 **The forecast concluded on the rear side of the copy form**

The telephone numbers to use for the voice-based AIRMET service are on the pilot's proforma in AIP GEN 3-5. Refer to the Get Met booklet for details on using facsimile equipment or the Internet to obtain the Area Forecasts for the three AIRMET regions.

```
AIRMET AREA FORECAST, NORTHERN REGION,
VALID DEC 24/1100Z TO 1900Z.

MET-SITUATION: AN UNSTABLE N TO NW AIRSTREAM, WILL COVER THE
               REGION.

               STRONG WIND WARNING: SFC WINDS WILL EXCEED 20KT, AT
               TIMES.

WINDS:
  1000FT:330/30KT PS01,  BEC 300/20KT PS01.
  3000FT:340/35KT MS03,  BEC 320/25KT MS02.
  6000FT:350/35KT MS09,  BEC 320/25KT MS09.
FREEZING LEVEL:2000FT.


WEATHER-CONDITIONS: 2 ZONES:

  ZONE 1: N OF A LINE, TEESSIDE, MANCHESTER, STRUMBLE, BELFAST.

          GEN 30KM, WITH 5/8CUSC 2500FT/6000.
          OCNL 6KM IN RA OR SLEET SH, WITH 6/8CU 1500FT/10000.
          ISOL 800M IN SN SH WITH TS, AND 6/8CB 500FT/20000.
    WRNG: CLD WILL COVER HILLS.
          MOD, ISOL SEV, TURB AND ICE IN CLD.
          MOD TURB BLW 6000FT.

  ZONE 2: ELSEWHERE.

          25KM, WITH 6/8CUSC 2000FT/6000.
          E OF 1 DEG W, 5/8AC 10000FT/18000, AND OCNL 6KM IN RA OR
          SLEET SH, WITH 6/8CU 1000FT/10000.
    WRNG: CLD WILL COVER HILLS.
          MOD TURB AND ICE IN CLD.
          IN E, MOD TURB BLW 6000FT.

OUTLOOK: UNTIL DEC 25/0100Z:

          SH BEC GEN ISOL, APART FROM NEAR IRISH SEA.
```

■ *Figure 26-4* **Example of a printed AIRMET Area Forecast (reduced) for the northern AIRMET region (obtained by facsimile)**

Graphic Area Forecasts

Graphic Area Forecasts are distributed by forecast offices and cover the entire UK on each chart. For VFR operations the *UK Low-Level Weather Forecast chart* (Metform 215), which shows general weather below approximately 15,000 ft, and the *UK Spot Wind chart,* which details spot winds and temperatures to 24,000 ft, are basic necessities. These are issued four times a day and cover a six-hour period.

An amended Metform 215 is issued with the word AMENDED at the top-left and bottom-left corners of the form. The example in Figure 26-5 is an amended chart.

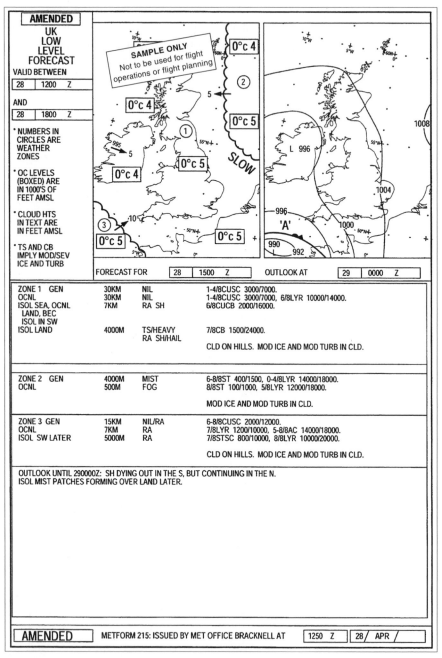

■ Figure 26-5 **A reduced sample of a UK Low-Level Forecast chart**
(obtained by using the dial-up METFAX service)

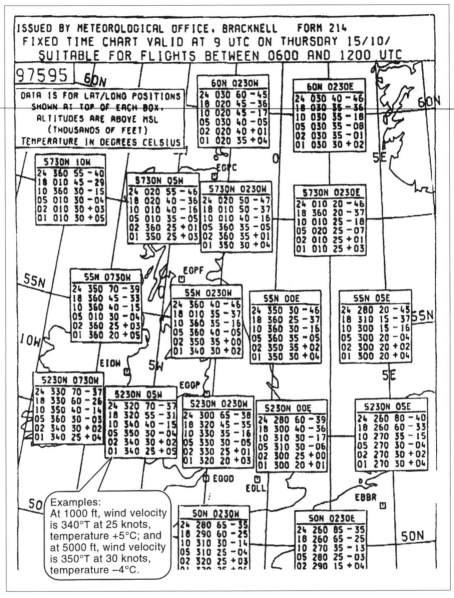

ISSUED BY METEOROLOGICAL OFFICE, BRACKNELL   FORM 214
FIXED TIME CHART VALID AT 9 UTC ON THURSDAY 15/10/
SUITABLE FOR FLIGHTS BETWEEN 0600 AND 1200 UTC

97595 60N

DATA IS FOR LAT/LONG POSITIONS
SHOWN AT TOP OF EACH BOX.
ALTITUDES ARE ABOVE MSL
(THOUSANDS OF FEET)
TEMPERATURE IN DEGREES CELSIUS

Examples:
At 1000 ft, wind velocity
is 340°T at 25 knots,
temperature +5°C; and
at 5000 ft, wind velocity
is 350°T at 30 knots,
temperature –4°C.

■ *Figure 26-6* **A reduced sample excerpt of a UK Spot Winds chart**

To decode weather forecasts and reports it is necessary to know the meaning of various symbols and abbreviations. These are included later in this chapter. Symbols can also be found in the AIP GEN 3-5 section, and abbreviations are included in the general list of abbreviations in AIP GEN 3-5.

Explanatory notes for the Low-Level Weather chart (METFORM 215T) are available from forecast offices and the dial-up METFAX Service. AIP GEN 3-5 also contains information on Form 215T.

Special Forecasts

For departure from aerodromes where the weather information is inadequate or unavailable, the pilot can request a forecast office to prepare a Special Forecast specifically for his flight. This takes time and so at least 2 hours' notice is required (4 hours' if the route distance exceeds 500 nm).

Special Forecasts are issued via facsimile, AFTN or telex; however, if your departure aerodrome is not so equipped, you can phone the forecast office for a dictation. Special Forecasts can also include Aerodrome Forecasts for the departure, destination and up to three alternates.

NOTE As Special Forecasts only cover that portion of the route outside the area of coverage of UK Area Forecasts, an appropriate UK Area Forecast must also be obtained. For further information consult AIP GEN 3-5.

Aerodrome Forecasts (TAFs)

Aerodrome Forecasts, known as TAFs, are text messages which follow the international (ICAO) format for aerodromes where observations are taken.

TAFs describe the forecast prevailing conditions at an aerodrome and usually cover periods of 9 to 24 hours.

The validity periods of many 24-hour TAFs do not start until 8 hours after the nominal time of origin and the forecast details only cover the *last* 18 hours.

The 9-hour TAFs are updated and re-issued every 3 hours; TAFs valid for 12–24 hours are updated and re-issued every 6 hours. Amendments are issued as and when necessary.

They may be preceded on a print-out by the term TAF, or, when on a list of one or more aerodromes, by FC or FT (forecast), followed by UK to represent the country: for instance 'FCUK'.

The TAF decode is included at the end of this chapter and can be found in AIP GEN 3-5. The four-figure ICAO location identifiers for UK aerodromes are listed in AIP GEN 2-4 and *Pooley's Flight Guide*.

NOTE Since 1 July 1999 all TAFs and METARs include the date and time of origin, e.g.

TAF EGDM **04 1330Z** 04 1524 120 15G25KT 9999 SCT 020

At the time of writing this change was not reflected in the PPL exam papers.

Aerodrome Forecast Examples

EXAMPLE 1

TAF EGPD 0716 13012KT 4000 OVC004=

This decodes to read:

Location: Aberdeen (Dyce)

~~Period of validity: 0700–1600 UTC~~

Surface wind: 130°T/12 knots

Visibility: 4,000 metres

Weather: nil

Cloud: sky covered with cloud, base 400 ft above aerodrome elevation

Significant variations: nil

NOTE The '=' sign after 004 indicates the end of the forecast message.

EXAMPLE 2

EGSS 1019 18020KT 9999 BKN022 TEMPO 1012 5000 BKN015=

This decodes to read:

Location: London (Stansted)

Period of validity: 1000–1900 UTC

Surface wind: 180°T/20 knots

Visibility: in excess of 10 km

Weather: nil

Cloud: 5–7 oktas (broken) layer, base 2,200 ft aal (cloud base in TAFs is above aerodrome level and not amsl)

Significant variations: temporary periods (TEMPO), i.e. less than 60 minutes, from 1000 until 1200 UTC, the visibility will be reduced to 5,000 metres with 5–7 oktas of cloud (broken), base 1,500 ft aal.

EXAMPLE 3

EGNV 1019 24025G35KT 8000 HZ SCT010 BKN025 BKN100 TEMPO 1012 1000 SHRA OVC005 PROB 30 TEMPO 1216 9999=

This decodes to read:

Location: Teesside

Period of validity: 1000–1900 UTC

Surface wind: 240°T/25 knots, with gusts to 35 knots

Visibility: 8,000 metres

Weather: haze

Cloud: 3–4 oktas (scattered), base 1,000 ft aal 5–7 oktas (broken), bases 2,500 ft and 10,000 ft aal.

Significant variations: for temporary periods (less than 60 minutes) between 1000 and 1200 UTC, the visibility will be reduced to 1,000 metres in moderate rain showers, with cloud 8 oktas (i.e. overcast), base 500 ft aal; plus a 30% probability for temporary periods between 1200 and 1600 UTC, visibility will be in excess of 10 kilometres.

Meteorological Reports

METARs

Metars are routine aerodrome reports. Routine weather observations are taken at many aerodromes on the hour and half-hour and issued on a routine basis, usually preceded by the abbreviations METAR or SA (surface actual). A code almost identical to that used for TAFs is used, with some variations:

☐ the time of observation is specified in a METAR as a four-figure group of hours and minutes past the hour (whereas TAFs have a four-figure group, the first two representing the hour of commencement of the validity period, and the final two the hour at which it ends);

☐ two temperatures are given: e.g. 09/07, where 09 is the actual temperature and 07 is the dewpoint temperature – the difference between them acting as a guide to the possibility of mist/fog that will occur should the temperature and the dewpoint become the same.

EXAMPLE 4

EGPE 0750 08009KT 3500 -RA SCT003 SCT005 BKN007 09/08 Q1006=

This METAR (actual aerodrome meteorological observation) decodes as:

Location: Inverness (Dalcross)

Time of report: 0750 UTC

Surface wind: 080°T at 9 knots

Visibility: 3,500 metres

Weather: light rain

Cloud: 3–4 oktas, bases 300 ft and 500 ft; 5–7 oktas, base 700 ft aal

Air temperature: +9°C

Dewpoint temperature: +8°C (i.e. if the temperature falls to this, mist or fog will form)

QNH: 1006 millibars.

Trends (or Landing Forecasts)

Trend forecasts are sometimes added to the end of METARs to forecast the weather changes expected to occur in the **two hours** immediately after the time of the observation. If no significant change is expected, the observation will be followed by the trend statement: NOSIG.

A Trend Forecast Attached to a METAR

EXAMPLE 5

METAR EGBO 0940 25015G28KT 3000 +RA BKN035 05/M02 Q1002 NOSIG=

This decodes as:

Actual weather for Halfpenny Green at 0940 UTC.

Wind: 250°T/15 knots gusting to 28 knots

Visibility: 3,000 metres in heavy rain

Cloud: 5–7 oktas (broken), base 3,500 ft aal

Temperature: 5°C; dewpoint: minus 2°C

QNH: 1002 millibars

No significant change forecast for the two hours to 1140 UTC.

Other more involved Trend Forecasts will often be preceded by BECMG (becoming) or TEMPO (temporary periods less than 60 minutes), with the time that the change is expected preceded by FM (from), TL (until) or AT (at). These terms are explained in more detail shortly.

VHF In-Flight Weather Reports

Weather information may be obtained at any time by radio from the Flight Information Service (FIS) or Air Traffic Control (ATC), who will also initiate a broadcast of any hazardous or significant weather that may be relevant to aircraft in the area (e.g. severe turbulence, thunderstorm activity, icing conditions, fog, etc.).

As mentioned previously, weather reports and trends for selected aerodromes are broadcast continuously on discrete VHF frequencies. This service is called **VOLMET**. The VOLMET broadcast for each aerodrome is updated each hour and half-hour and includes:

- the actual weather report;
- landing forecast;
- SIGMET (significant weather, if any); and
- the forecast trend for the two hours following the time of the report.

Information is also obtainable from the **Automatic Terminal Information Service (ATIS)** − a tape-recorded message of the current aerodrome information and is broadcast on appropriate VOR or discrete VHF frequencies to off-load the ATC VHF communications frequencies.

Some aerodromes have both an arrivals and departure ATIS. One variation on the ATIS is that the wind direction is given in degrees magnetic to allow the pilot to relate it to the runway direction (which is in °M) more easily. This also applies to winds passed to the pilot by the tower.

Cloud Bases

- The cloud base in a TAF (Aerodrome Forecast) or a Trend Forecast (both of which refer to a particular aerodrome), is given **above aerodrome level (aal)**; whereas
- Area Forecasts give the cloud base **above mean sea level (amsl)**, i.e. as altitudes.

TAFs and Trend Forecasts, then, provide the pilot with an immediate appreciation of the cloud ceiling at a particular aerodrome; 'aal' is, in fact, the height above the highest point in the landing area.

Over large areas of countryside, however, with mountains, valleys, plateaux and coastlines, a constant reference for cloud height is needed; of course the only satisfactory one is mean sea level (msl). To determine the expected level of the cloud base above the ground from the information in an Area or Regional Forecast you need to know the ground elevation. This can be determined from a suitable aeronautical chart.

Symbols and Abbreviations

Symbols on UK Low-Level Weather Charts

Graphic Area Forecasts and Special Forecasts will contain symbols and abbreviations from the following standard sets.

1   Symbols for significant Weather, Tropopause and Freezing Level, etc.

Symbol	Meaning	Symbol	Meaning
Thunderstorm	Thunderstorm	Rain	Rain
Tropical cyclone	Tropical cyclone	Snow	Snow
Severe squall line	Severe squall line	Widespread blowing snow	Widespread blowing snow
Hail	Hail	Shower	Shower
Moderate turbulence	Moderate turbulence	Severe sand or dust haze	Severe sand or dust haze
Severe turbulence	Severe turbulence	Widespread sandstorm or duststorm	Widespread sandstorm or duststorm
Marked mountain waves	Marked mountain waves	Widespread haze	Widespread haze
Light aircraft icing	Light aircraft icing	Widespread mist	Widespread mist
Moderate aircraft icing	Moderate aircraft icing	Widespread fog	Widespread fog
Severe aircraft icing	Severe aircraft icing	Freezing fog	Freezing fog
Freezing precipitation	Freezing precipitation	Widespread smoke	Widespread smoke
Drizzle	Drizzle	Volcanic eruption	Volcanic eruption

Note: Altitudes between which phenomena and any associated cloud are expected are indicated by flight levels, top over base or top followed by base. 'XXX' means the phenomenon is expected to continue above and/or below the vertical coverage of the chart. Phenomena of relatively lesser significance, for example light aircraft icing or drizzle, are not usually shown on charts even when the phenomenon is expected. The thunderstorm symbol implies hail, moderate or severe icing and/or turbulence.

400 Tropopause spot altitude (eg FL400)

~~~~~~ Boundary of area of significant weather

H 440 High point or maximum in tropopause topography (eg FL440)

– – – – Boundary of area of clear air turbulence. The CAT area may be marked by a numeral inside a square and a legend describing the numbered CAT area may be entered in the margin

340 L Low point or minimum in tropopause topography (eg FL340)

10 State of sea (wave height in metres)

0°:100 Freezing level

18 Sea surface temperature (°C)

■ *Figure 26-7* **Symbols for significant weather on MET charts**

Altitudes between which phenomena are expected are indicated by flight levels, top over base. 'XXX' means the phenomenon is expected to continue above and/or below the vertical coverage of the chart.

Phenomena of relatively lesser significance – for example, light aircraft icing or drizzle – are not usually shown on charts even when the phenomena are expected. The thunderstorm symbol, for example, implies hail, moderate or severe icing and/or turbulence.

2	**Fronts and Convergence Zones**		
▲▲ Cold front at the surface		———	Axis of trough
●● Warm front at the surface		∧∧∧∧∧	Axis of ridge
▲●▲▲ Occluded front at the surface		⟩⟩⟩	Convergence line
●●▼ Quasi-stationary front at the surface		⫿⫿	Inter-tropical convergence zone

Note: An arrow with associated figures indicates the direction and the speed of the movement of the front (knots). Dots inserted at intervals along the line of a front indicate it is a developing feature (frontogenesis), while bars indicate it is a weakening feature (frontolysis).

■ *Figure 26-8* **Symbols for fronts and convergence zones on met charts**

NOTE An arrow with associated figures indicates the direction and speed of movement of the front in knots.

Weather Abbreviations

Cloud types are listed according to the standard set of abbreviations:

CLOUD TYPE ABBREVIATIONS	
CI	*Cirrus*
CC	*Cirrocumulus*
CS	*Cirrostratus*
AC	*Altocumulus*
AS	*Altostratus*
NS	*Nimbostratus*
SC	*Stratocumulus*
ST	*Stratus*
CU	*Cumulus*
CB*	*Cumulonimbus* **CB implies hail, moderate or severe icing and/or turbulence*
LYR	*Layer or layered (instead of cloud type)*

Altitudes are indicated in flight levels, top over base. For example 30/220 is base 3,000 ft, tops at 22,000 ft.

The amount of **cloud coverage** in Area Forecasts (and in TAFs and METARs in countries such as Australia) is expressed in *eighths* of the sky covered, or *oktas*. 4 oktas means that half of the sky is covered by the cloud mentioned, whereas 8 oktas means complete cloud coverage.

In Aerodrome Forecasts and Reports, the amount of cloud coverage is indicated by the following abbreviations:

CLOUD AMOUNT ABBREVIATIONS (EXCEPT CB)	
SKC	Sky clear (0 oktas)
FEW	Few (1–2 oktas)
SCT	Scattered (3–4 oktas)
BKN	Broken (5–7 oktas)
OVC	Overcast (8 oktas)

Visual navigation with reference to the ground or water is difficult above more than 4 oktas of cloud (half the sky covered). If more than 4 oktas (or two lots of 1–4 oktas) appear on the Area Forecast, you should compare the forecast cloud base with the elevation of the en route terrain and consider carefully whether or not safe navigation to your destination is possible beneath the clouds.

OVC
(overcast)

BKN
(broken)

SCT
(scattered)

FEW
(few)

■ Figure 26-9 **Cloud amount abbreviations**

Think about how much of the ground you would actually see if you were navigating en route above 4 oktas of cloud predicted in an Area Forecast.

■ Figure 26-10 **4CU035 severely restricts the pilot's view of the ground**

Thunderstorms

Thunderstorms (TS), which are best avoided by aircraft, are associated with cumulonimbus (CB) clouds. The amount of CB cloud in an area is indicated by the following abbreviations:

CB AMOUNT ABBREVIATIONS	
ISOL	*Isolated – for individual CB clouds*
OCNL	*Occasional – for well-separated CB clouds*
FRQ	*Frequent – for CB clouds with little or no separation*
EMBD	*Embedded – CB clouds contained in layers of other clouds*

  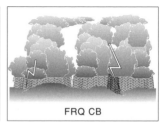

ISOL CB   OCNL CB   FRQ CB

■ *Figure 26-11* **Cumulonimbus amount variations**

CAUTION TS and CB each imply hail, moderate or severe icing and/or turbulence.

Descriptive Abbreviations

In addition to the abbreviations already covered, the following main weather abbreviations may be found in text forecasts and reports, including AIRMET regional forecasts obtained from the METFAX Service or in an aerodrome briefing office. (AIP GEN 3-5 has an extensive abbreviations list, including those which apply to meteorological forecasts and reports.)

COMMON MET ABBREVIATIONS					
+	*heavy*	**ICE**	*icing*	**RA**	*rain*
–	*light*	**IMPR**	*improve(-ing)*	**RAG**	*ragged*
BECMG	*becoming*	**IMT**	*immediately*	**RE**	*recent*
BL	*blowing*	**INC**	*in cloud*	**SEV**	*severe*
BLO	*below clouds*	**INCR**	*increase*	**SFC**	*surface*
BR	*mist*	**INTSF**	*intensify*	**SG**	*snow grains*
BTN	*between*	**IR**	*ice on runway*	**SH**	*showers*
CAT	*clear air turbulence*	**LAN**	*inland*	**SN**	*snow*
COT	*at the coast*	**LGT**	*light*	**SQ**	*squall*

COMMON MET ABBREVIATIONS					
CUF	cumuliform	**LOC**	locally	**STNR**	stationary
DECR	decrease	**LSQ**	line squall	**TCU**	towering cumulus
DP	dewpoint temperature	**LV**	light and variable (wind)	**TS**	thunderstorm
DR	low drifting	**LYR**	layer(ed)	**TURB**	turbulence
DTRT	deteriorate	**MI**	shallow	**UA**	air report (PIREP)
DZ	drizzle	**MNM**	minimum	**V**	varying
FCST	forecast	**MOD**	moderate	**VAL**	in valleys
FG	fog	**MS**	minus	**VC**	vicinity of aerodrome
FLUC	fluctuating	**MT**	mountain	**VCY**	vicinity
FT	feet	**MTW**	mountain waves	**VER**	vertical
FU	smoke	**NC**	no change	**VIS**	visibility
FZ	freezing	**NSC**	nil significant cloud	**WDSPR**	widespread
G	gusting	**NSW**	nil significant weather	**WKN**	weaken(ing)
GEN	generally	**OBSC**	obscured	**WRNG**	warning
GR	hail	**PE**	ice pellets	**WS**	windshear
GS	small hail and/or snow pellets	**PROV**	provisional	**WX**	weather
HVY	heavy	**PS**	plus		

CAVOK

The term CAVOK is used frequently and you should know its definition. CAVOK means that the following conditions occur simultaneously:

☐ visibility 10 km or more;

☐ no cloud below 5,000 ft above aerodrome level (aal) or below the highest minimum sector altitude, whichever is the higher, and no cumulonimbus;

☐ no significant weather phenomena at or in the vicinity of the aerodrome.

CAUTION Do not fall into the trap (as some VFR pilots have), of thinking that CAVOK means sky clear – CAVOK does not necessarily mean blue skies. With CAVOK you could have complete cloud coverage above 5,000 ft, and any non-IMC or non-instrument-rated pilot cruising along at, say, FL90 might have trouble getting down through this cloud.

Also, a complication with the term CAVOK, which can affect non-IMC and non-instrument-rated pilots, is that the criteria regarding changes in cloud amount and base for amending the TAF apply only when there are significant changes below 1,500 ft aal. This could mean that a forecast containing CAVOK may not be amended until the cloud base falls below 1,500 ft.

Perhaps the safest interpretation of CAVOK, even though it specifically refers to no cloud below 5,000 ft aal, is to assume that it really means that the cloud base should not fall below 1,500 ft for the period of the forecast.

Changing Weather in Forecasts

Temporary Change (TEMPO)

While the cumulonimbus cloud associated with a thunderstorm may exist for hours, its passage through the immediate vicinity of an aerodrome may take only a brief period – less than 60 minutes, or an even shorter time. During these *temporary* periods the weather in the vicinity of the aerodrome might be quite different when compared with the background of the prevailing weather.

In such a situation, the Aerodrome and Landing Forecasts would state the general conditions existing at the aerodrome (i.e. the **prevailing conditions**), and **any temporary changes** to the conditions would be indicated by the term TEMPO.

TEMPO. Temporary variation lasting less than 60 minutes or, if recurring, lasting in total less than half the TREND (or TAF, see AIP GEN 3-5) period; that is, changes take place sufficiently infrequently for the prevailing conditions to remain those forecast for the period.

EXAMPLE 6

TAF EGBS 2008 12010KT 9999 DZ SCT010TCU BKN020 TEMPO 2023 5000 BKN010CB +SHRAGR=

We interpret this to mean:
- the Aerodrome Forecast (TAF) is for Shobdon
- for the period 2000 to 0800 UTC

prevailing conditions:
- wind velocity 120°T/10 knots
- visibility in excess of 10 km (indicated by 9999)
- drizzle
- cloud: 3–4 oktas of towering cumulus at base 1,000 ft aal; 5–7 oktas of other cloud at 2,000 ft aal (CB and TCU are the only cloud types shown on TAFs and METARs)

with periods of less than 60 minutes (TEMPO):
- between 2000 and 2300 UTC, with
- visibility reduced to 5,000 metres, and
- 5–7 oktas of cumulonimbus at base 1,000 ft aal★
- heavy rain showers with hail.

★ In other words, temporary deteriorations in the Shobdon weather during the three hours from 2000 to 2300 UTC; conditions as specified.

NOTE TEMPO can relate to improvements as well as deteriorations in wind, visibility, weather or cloud.

Lasting Changes

Whereas TEMPO is used to indicate a temporary variation from the prevailing weather, when lasting changes in the prevailing weather are forecast, the term **BECMG** (becoming) is used preceding an expected permanent change in the weather conditions.

In an Aerodrome Forecast (TAF), BECMG will be followed by a four-figure time group; in a Trend Forecast that is appended to an aerodrome report (METAR), BECMG may be followed by a four-figure time group in hours and minutes preceded by one of the abbreviations: **FM** (from), **TL** (until) or **AT** (at).

Once the BECMG changes are completed, there is new prevailing weather. Once TEMPO events are finished, however, the original prevailing weather re-asserts itself.

EXAMPLE 7

TAF EGDL 0716 08008KT 9999 BKN008 BECMG 1012 9999 BKN020 BECMG 1315 CAVOK

We interpret this Aerodrome Forecast (TAF) for Lyneham aerodrome, valid from 0700 to 1600 UTC, to mean:

initial prevailing weather:
- wind from 080°T at 8 knots
- visibility in excess of 10 km
- 5–7 oktas at 800 ft aal, plus:

  becoming new prevailing conditions over a two-hour period commencing around 1000 UTC and completed by around 1200 UTC:
- visibility in excess of 10 kilometres; cloud: 5–7 oktas at 2,000 ft aal

  (this should be the weather at 1200 hours lasting until the next forecast change), plus:

  becoming, between 1300 and 1500 UTC, CAVOK.

EXAMPLE 8

FCUK EGBN 0716 33018G35KT 3600 +RA BKN008 BKN020 OVC070 MOD TURB BLW 6000 FT BECMG 1314 16015KT 9999 NSW BKN020 SCT100

9-hour Aerodrome Forecast (TAF) for Nottingham, valid 0700 to 1600 UTC

initial prevailing weather:
- a gusty wind of 18 to 35 knots from the north-west (330°T);
- visibility reduced to 3,600 metres in heavy rain

- cloud: 5–7 oktas at 800 ft and at 2,000 ft aal, and 8 oktas at 7,000 ft aal
- moderate turbulence below 6,000 ft, plus

becoming, between 1300 and 1400 UTC (a fairly rapid change):
- a southerly wind of 15 knots (from 160°T)
- visibility in excess of 10 kilometres
- an improvement to no significant weather
- cloud: 5–7 oktas at 2,000 ft aal, and 3–4 oktas at 10,000 ft aal

In other words, the forecaster is expecting quite a significant improvement in the weather.

EXAMPLE 9

SAUK EGSS 220V28025KT 8000 RERA SCT006 BKN012 10/09 Q0988 BECMG AT 1300 CAVOK

This is an aerodrome observation (METAR) for Stansted aerodrome:
- wind varying between 220°T and 280°T at 25 knots
- visibility 8,000 metres
- recent rain
- cloud: 3–4 oktas at 600 ft aal; 5–7 oktas at 1,200 ft aal
- temperature: 10°C and dewpoint 09°C (close to mist or fog conditions)
- QNH: 988 millibars (quite low).

Trend: becoming, at 1300 UTC, CAVOK conditions.

Other descriptive terms that appear on Trend statements are: **NOSIG:** no significant change, **NSC:** no significant cloud, and **NSW:** no significant weather.

Probability

Sometimes the forecaster is uncertain of whether conditions will occur and, if he or she assesses the probability of them occurring as 40% or less, the message may be prefaced with a PROB (probability) percentage.

EXAMPLE 10

PROB 40 TEMPO 1012 1000 +TS BKN005 BKN040CB

If the above was included in an Aerodrome Forecast (TAF), it means that the forecaster has placed a **40% probability factor** of periods of less than 60 minutes' duration (TEMPO), between the hours of 1000 and 1200 UTC, with:
- 1,000 metres visibility
- severe thunderstorm activity
- 5–7 oktas at 500 ft aal
- 5–7 oktas cumulonimbus with a base of 4,000 ft aal.

What if Poor Weather is Forecast?

While most of your cross-country navigational flights will be conducted in reasonably good weather conditions, occasionally you may encounter conditions at the met briefing (either forecast or actual) that are not so good. Then you are faced with a difficult decision – does the flight commence and if so under what conditions?

Like most things in flying, it is not the everyday situation that tests us, but rather the unusual situations that have a habit of occurring from time to time. Making sound operational decisions is what flying is all about.

The following extracts from AIP GEN 3-5 contain more useful information about weather forecasts and reports.

Now complete **Exercises 26 – Weather Forecasts and Reports.**

---

1   **Aerodrome Weather Report Codes (Actuals)**

1.1   The content and format of an actual weather report is as shown in the following table:

Report Type	Location Identifier	Date/Time	Wind	Visibility	RVR
METAR	EGSS	231020Z	31015G30KT 280V350	1400SW 6000N	R24/P1500

Present WX	Cloud	Temp/Dew Pt	QNH	Recent WX	WindShear	TREND	Rwy State
SHRA	FEW005 SCT010CB BKN025	10/03	Q0995	RETS	WS RWY23	NOSIG	88290592

1.1.1   **Identifier**

The identifier has three components:

(a) Report type
  (i) METAR - Aviation routine weather report. These are compiled half-hourly or hourly at fixed times while the aeronautical station is open;
  (ii) SPECI - Aviation selected special weather report. Special reports are prepared to supplement routine reports when improvements or deteriorations through certain criteria occur. However, by ICAO Regional Air Navigation agreement, they are not disseminated by either OPMET in UK or MOTNE in Europe.

(b) Location indicator
  ICAO four-letter code letters (for UK aerodromes).

(c) Date/Time
  The date and time of observation in hours and minutes UTC, followed by the letter Z.

  Example: METAR EGSS 231020Z.

Note: In the case of a meteorological bulletin which may consist of reports from one or more aerodromes, the code name METAR (or SPECI) may be replaced by the abbreviation **SA** (or **SP**) followed by the bulletin identifier and the date and time of the observation. In the UK, usually neither code name nor time will appear in the individual reports.

1.1.2   **Wind**

1.1.2.1   Wind direction is given in degrees True (three digits) rounded to the nearest 10 degrees, followed by the windspeed (two digits, exceptionally three), both usually meaned over the ten minute period immediately preceding the time of observation. These are followed without a space by one of the abbreviations **KT**, **KMH** or **MPS**, to specify the unit used for reporting the windspeed:

  Example: 31015KT.

1.1.2.2   A further two or three digits preceded by a G gives the maximum gust speed in knots when it exceeds the mean speed by 10 kt or more:

  Example: 31015G30KT.

1.1.2.3   Calm is indicated by '00000', followed by the units abbreviation, and variable wind direction by the abbreviation **'VRB'** followed by the speed and unit.

1.1.2.4   If, during the 10 minute period preceding the time of the observation, the total variation in wind direction is 60° or more, the observed two extreme directions between which the wind has varied will be given in clockwise order, separated by the indicator letter V but only when the speed is greater than 3kt:

Example: 31015G30KT 280V350.

1.1.3   Horizontal Visibility

1.1.3.1   When there is no marked variation in visibility by direction the minimum is given in metres. When there is a marked directional variation in the visibility, the reported minimum will be followed by one of the eight points of the compass to indicate the direction:

Example: 4000NE.

1.1.3.2   When the minimum visibility is less than 1500 metres and the visibility in another direction greater than 5000 metres, additionally the maximum visibility and its direction will be given:

Example: 1400SW 6000N.

Note: 9999 indicates a visibility of 10 km or more; 0000 a visibility of less than 50 metres.

1.1.5   Weather

1.1.5.1   Each weather group may consist of appropriate intensity indicators and letter abbreviations combined in groups of two to nine characters and drawn from the following table:

Significant Present and Forecast Weather Codes

Qualifier		Weather Phenomena		
Intensity or Proximity	Descriptor	Precipitation	Obscuration	Other
- Light	MI — Shallow	DZ — Drizzle	BR — Mist	PO — Dust/Sand Whirls (Dust Devils)
Moderate (no qualifier)	BC — Patches	RA — Rain	FG — Fog	
	BL — Blowing	SN — Snow	FU — Smoke	SQ — Squall
+ Heavy ('Well developed' in the case of FC and PO)	SH — Shower(s)	SG — Snow Grains	VA — Volcanic Ash	FC — Funnel Cloud(s) (tornado or water-spout)
	TS — Thunderstorm	IC — Ice Crystals (Diamond Dust)	DU — Widespread Dust	
	FZ — Freezing (Super-Cooled)	PE — Ice-Pellets	SA — Sand	SS — Sandstorm
VC In the vicinity (not at the aerodrome but not further away than approx 8 km from the aerodrome perimeter)	PR — Partial (covering part of aerodrome)	GR — Hail	HZ — Haze	DS — Duststorm
		GS — Small hail (<5 mm diameter) and/or snow pellets		

1.1.5.2   Mixture of precipitation types may be reported in combination as one group, but up to three separate groups may be inserted to indicate the presence of more than one independent weather type:

Examples: MIFG, VCSH, +SHRA, RASN, -DZ HZ.

Note 1:   BR, HZ, FU, IC, DU and SA will **not** be reported when the visibility is greater than 5000 m.

Note 2:   Some codes are shown that will not be used in UK METARs and TAFs but may be seen in continental reports and when flying in Europe.

2   Aerodrome Forecast (TAF) Codes

2.1   TAFs describe the forecast prevailing conditions at an aerodrome and usually cover a period of 9 to 24 hours. The validity periods of many of the latter do not start until 8 hours after the nominal time of origin and the forecast details only cover the last 18 hours. The 9 hour TAFs are updated and re-issued every 3 hours and those valid for 12 to 24 hours, every 6 hours. Amendments are issued as and when necessary. The forecast period of a TAF may be divided into two or more self contained parts by the use of the abbreviation FM followed by a time. TAFs are issued separately from the METAR or SPECI and do not refer to any specific report; however, many of the METAR groups are also used in TAFs and significant differences are detailed below.

2.2   The content and format of a TAF is as in the following table:

Report Type	Location Identifier	Date/Time of Origin	Validity Time	Wind	Visibility	Weather
TAF	EGZZ	130600Z	130716	31015KT	8000	-SHRA

Cloud		Variant	Validity Times
FEW005 SCT018CB BKN025		TEMPO	1116

Visibility	Weather	Cloud	Probability	Validity Time	Weather
4000	+SHRA	BKN010CB	PROB30	1416	TSRA

2.2.1 Differences from the METAR

(a) **Identifier.** In the validity period, the first two digits indicate the day on which the period begins, the next two digits indicate the time of commencement of the forecast in whole hours UTC and the last two digits are the time of ending of the forecast in whole hours:

Note: In the case of a TAF bulletin which may consist of forecasts for one or more aerodromes, the code name TAF will be replaced by the abbreviation FC or FT, followed by a bulletin identifier and the date and time of origin (hours and minutes UTC), and neither code name nor date/time will appear in the individual forecasts.

(b) **Horizontal Visibility.** As with the METAR code, except that only one value (the minimum) will be forecast.

(c) **Weather.** If no significant weather is expected the group is omitted. However, after a change group, if the weather ceases to be significant, the abbreviation NSW is used for No Significant Weather.

(d) **Cloud.** When clear sky is forecast the cloud group is replaced by SKC (sky clear). When no cumulo-nimbus or clouds below 5000 ft or below the highest minimum sector altitude, whichever is the greater, are forecast and CAVOK or SKC are not appropriate, then NSC (No Significant Cloud) is used. Only CB cloud will be specified.

(e) **Significant Changes.** The abbreviation FM followed by the time to the nearest hour and minute UTC is used to indicate the beginning of a self contained part in a forecast. All conditions given before this group are superseded by the conditions indicated after the group:

Example: FM1220 27017KT 4000 BKN010.

The change indicator BECMG followed by a four figure time group, indicates an expected permanent change in the forecast meteorological conditions, at either a regular or irregular rate, occurring at an unspecified time within the period:

Example: BECMG 2124 1500 BR.

The change indicator TEMPO followed by a four figure time group indicates a period of temporary fluctuations to the forecast meteorological conditions which may occur at any time during the period given. The conditions following these groups are expected to last less than one hour in each instance and in aggregate less than half the period indicated:

Example: TEMPO 1116 4000 +SHRA BKN010CB.

(f) **Probability.** The probability of occurrence happening will be given as a percentage, although only 30% and 40% will be used. The abbreviation PROB is used to introduce the group, followed by a time group, or an indicator and a time group:

Examples: (a) PROB30 0507 0800 FG BKN004: (b) PROB40 TEMPO 1416 TSRA.

(g) **Amendments.** When a TAF requires amendment, the amended forecast may be indicated by inserting AMD after TAF in the identifier and this new forecast covers the remaining validity period of the original TAF:

Example: TAF EGZZ AMD 130820Z 130816.

If an amended TAF bulletin is issued, amendment indicators 'AAA', or 'AAB' etc, and the time of origin will appear in the header line beginning FC or FT; in UK usually neither 'AMD' nor the date/time will appear in the TAF.

'CCA', 'CCB' may also be seen for corrections and 'RRA', 'RRB' for retards.

5 Wind Symbols

5.1 Wind/Temperature Chart

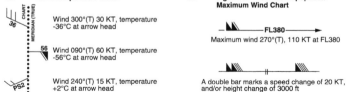

Wind 300°(T) 30 KT, temperature -36°C at arrow head

Wind 090°(T) 60 KT, temperature -56°C at arrow head

Wind 240°(T) 15 KT, temperature +2°C at arrow head

5.2 Significant Weather/Tropopause/ Maximum Wind Chart

FL380

Maximum wind 270°(T), 110 KT at FL380

A double bar marks a speed change of 20 KT, and/or height change of 3000 ft

6 UK Low Level Weather Chart – Metform 215

6.1 The Metform 215 is a forecast of in-flight conditions from the surface to around 15000 ft, covering the UK and near Continent. The form comprises a **fixed time forecast weather chart** and associated descriptive text covering the period of validity from 3 hours before the fixed time, to 3 hours after. A separate **outlook chart** shows the expected position of the principle synoptic features at the end of the outlook period of 6 hours, with separate text describing the main weather developments during this period.

6.2 **Information on Form.** The following sub-paragraphs summarise the contents of Metform 216 (Explanatory Notes for Form 215), available in A4 or larger size on application from METFAX by dialling 0336-400-505.

6.2.1 Main Forecast Weather Chart and Text

6.2.1.1 The fixed time weather chart at the top left of the form shows the forecast position, direction and speed of movement of surface fronts and pressure centres for the fixed time shown in the chart legend. The position of highs (H) and lows (L), with pressure values in millibars is shown by the symbols O and X. The direction and speed of movement (in knots) of fronts and other features is given by arrows and figures. Speeds of less than 5 knots are shown as 'SLOW'. All features are given identifying letters to enable their subsequent movements to be followed on the outlook chart.

6.2.1.2 Freezing levels (0°C) are shown in boxes as thousands of feet at appropriate places on the chart.

6.2.1.3 Zones of distinct weather patterns are enclosed by continuous scalloped lines, each zone being identified by a number within a circle. The forecast weather conditions (visibility, weather, cloud) during the period of validity, together with warnings and any remarks are given in the text underneath the charts, each zone being dealt with separately and completely.

6.2.1.4 In the text, surface visibility is expressed in metres (m) or kilometres (km), with the change over at 5 km.

Weather is described in plain language, using well known and self evident abbreviations. Cloud amount (in oktas) and type, with the height of base and top, is given, with all heights in feet amsl.

Warnings and significant changes and the expected occurrence of icing and turbulence are given in plain language, using standard abbreviations where possible.

The height of any sub-zero layer below the main layer is given in the text.

Hill fog is not used but 'cloud covering hills' is thought to be more informative and implies visibility < 200 m.

6.2.1.5 Single numerical values given for any element represent the most probable mean in a range of values, covering approximately ±25%.

AERODROME ACTUAL WEATHER – METAR AND SPECI DECODE

	Code Element	Example	Decode	Notes/Alternative Coding
1	**IDENTIFICATION*** a. METAR or SPECI b. Location Indicator c. Date/Time	METAR EGLL 291020Z	'METAR' 'London Heathrow' 'at ten twenty zulu on 29th'	METAR – aviation routine report, SPECI – selected special (not disseminated from UK civil aerodromes) Station 4 letter ICAO indicator Usually omitted when METARs are presented in a bulletin
2	**WIND** a. Wind direction/speed* b. Extreme direction variance	31015G27KT 280V350	'three one zero degrees, fifteen knots, max twenty seven knots' 'varying between two eight zero and three five zero degrees'	Max only given if ≥ 10 kt than mean. VRB = Variable 00000 kt = Calm Variation given in clockwise direction but only when mean speed >3 kt
3	**VISIBILITY** a. Minimum Visibility* b. Maximum Visibility	1400SW 6000N	'one thousand four hundred metres to the south west' 'six thousand metres to the north'	0000 = "less than 50 metres" 9999 = "ten kilometres or more" Direction of min visibility given by eight points of compass when required Given when min visibility <1500 m and max visibility > 5000 m
4	**RVR**	R27R/1100	'RVR, runway two seven right one thousand one hundred metres'	RVR tendency (U = increasing, D = decreasing, N = No Change) may be added after fig. i.e. R27R/1100D P1500 = more than 1500 m, M0050 = less than 50m Significant variations – example R24/0750V1100D i.e. varying between two values

5	**PRESENT WEATHER**	+SHRA	'heavy rain showers'	+ = Heavy (well developed in the case of +FC and +PO): – = Light; no qualifier = Moderate
				BC = Patches FZ = Freezing IC = Ice Crystals RA = Rain SQ = Squalls
				BL = Blowing DZ = Drizzle GR = Hail (≥ 5 mm) SA = Sand SS = Sandstorm
				BR = Mist FC = Funnel Cloud GS = Small Hail or SH = Showers TS = Thunderstorm
				DR = Drifting FG = Fog snow pellets SG = Snow grains VA = Volcanic Ash
				DS = Duststorm FU = Smoke PE = Ice Pellets SN = Snow VC = In Vicinity
				                    PO = Dust Devils
				                    HZ = Haze
				Up to three groups may be present, constructed by selecting and combining from the above. Group omitted if no weather to report

6	**CLOUD***	FEW005 SCT010CB BKN025	'Few at 500 feet Scattered cumulonimbus at one thousand feet. Broken at two thousand five hundred feet'	SKC = "Sky clear" (0 oktas) FEW = "Few" (1-2 oktas) SCT = "Scattered" (3-4 oktas) BKN = "Broken" (5-7 oktas) OVC = "Overcast" (8 oktas) Only two cloud types reported: TCU = Towering Cumulus CB = Cumulonimbus VV/// = "state of sky obscured" (cloud base not discernible): figures in lieu of ///give vertical visibility in hundreds of feet. Usually up to three, but occasionally more, cloud groups may be reported.
7	**CAVOK¹**	CAVOK	'Cav-oh-kay'	Vis ≥ 10km, no cumulonimbus cloud and no cloud below 5000 ft or highest MSA (greater), no weather significant to aviation
8	**TEMP/DEW POINT***	10/03	'Temperature ten degrees Celsius, Dew Point three degrees celsius'	If dew point missing, example temperature would be reported as 10///
9	**QNH***	Q0995	'nine nine five'	Q indicates millibars. If letter A is used, QNH is in inches and hundredths.
10	**RECENT WEATHER**	RETS	'Recent Thunderstorms'	RE = Recent, weather codes given above; up to three groups may be present.
11	**WINDSHEAR**	WS RWY24	'Windshear runway two four'	Will not be reported at present for UK aerodromes.
12	**TREND**	BECMG FM 1100 2503G50KT TEMPO FM0830 TL0830 3000 SHRA	'Becoming from 1100, 250° 35 knots max 50 knots, temporarily from 0630 until 0830 3000 metres. Moderate rain showers'	BECMG = Becoming TEMPO = temporarily NOSIG = No sig change NSW = No sig weather AT = At FM = From Any of the wind forecast, visibility, weather or cloud groups may be used, and CAVOK. TL = Until NSC = No sig cloud Multiple groups may be present

* Indicates a mandatory code element   ¹ CAVOK will replace visibility and cloud groups.

Examples of METARs in Code as they would appear in a Bulletin

SAUK02 EGGY 301220
METAR
EGLL 240015KT 200V280 8000 – RA FEW010 BKN025 OVC080 18/15 Q0983 TEMPO 3000 RA BKN 008 OVC020 =
EGPZ 240015G37KT 270V360 1200NE 6000S + SHSN SCT005 BKN10CB 03/M01 Q0999 RETS BECMG AT1300 NSW SCT015 BKN100 =
The above METARs for 1220 UTC on the 30th of the month in plain language:

EGLY: Surface wind: mean 240 deg true, 15 kt; varying between 200 and 280 deg; minimum vis 8 km; light rain; cloud: 1-2 oktas base 1000 ft, 5-7 oktas 2500ft, 8 oktas 8000ft; Temperature +18°C; Dew Point 15°C; QNH 983 mb; Trend: temporarily 3000 m in moderate rain with 5-7 oktas 800 ft, 8 oktas 2000 ft.

EGPZ: Surface wind: mean 300 deg true, 25 kt; maximum 37 kt, varying between 270 and 360 deg; minimum vis 6 km(to northeast), maximum vis 6 km(to south); heavy shower of snow, Cloud: 3-4 oktas base 500 ft, 5-7 oktas CB base 1000 ft; Temperature +3°C; Dew Point -1°C; QNH 999 mb; thunderstorm since previous report; Trend: improving at 1300UTC no sig weather, 3-4 oktas 1500 ft, 5-7 oktas 1000ft.

AERODROME FORECAST – TAF DECODE

	Code Element	Example	Decode	Notes/Alternative Coding
1	REPORT TYPE*	TAF	'TAF'	Name for an aerodrome forecast
2	LOCATION*	EGSS	'London Stansted'	Station 4 letter ICAO indicator
3	DATE/TIME OF ORIGIN	130600Z	'For 13th at oh six hundred Zulu'	Usually omitted
4	VALIDITY TIME	130716	'Valid from oh seven hundred to sixteen hundred on 13th'	Zulu
5	WIND*	31015G30KT	'Three zero degrees fifteen max thirty knots'	VRB = Variable 0000 kt = Calm
6	MIN VISIBILITY or CAVOK*	8000	'Eight kilometres'	9999 = 10km or more; 0000 = less than 50 metres
7	SIGNIFICANT WEATHER	–SHRA	'Light rain shower'	See present weather table on METAR page for details; NSW = No significant weather
8	CLOUD*	FEW005 SCT010 / SCT018CB BKN025	'Few at five hundred feet, scattered at one thousand feet, scattered cumulonimbus at one thousand eight hundred feet, broken at two thousand five hundred feet'	SKC = Sky clear FEW = Few 1-2 oktas SCT = 3-4 oktas BKN = 5-7 oktas; OVC = 8 oktas; VV/// = state of sky obscured (figures in lieu of /// give forecast vertical visibility in hundreds of feet); NSC = No significant cloud (none below 5000ft & no CB) CB will be the only cloud type specified.
9	SIGNIFICANT CHANGES			
a.	Probability	PROB30	'PROB thirty'	Normally only 30% or 40% Probability should be used. TEMPO may or may not be present.
b.	Time	1416	'From fourteen hundred to sixteen hundred'	Indicates beginning and end of forecast period in Co-ordinated Universal Time (UTC), (Zulu time (Z))
c.	Change indicator	a. BECMG 1416	'becoming from fourteen hundred to sixteen hundred' or	Also TEMPO = Temporarily may be used
d.	Met Groups	b. FM1400 / TSRA BKN010CB	'From fourteen hundred' followed by 'Thunderstorm with rain, broken cumulonimbus at one thousand feet'	Met group follows indicating a change in some or all of the elements forecast in the first part of the TAF

* Indicates a mandatory code element; CAVOK will replace visibility and cloud groups.

EXAMPLES

1BHR TAF

FTUK31 EGGY 102200
EGLL 110624 13010KT 9000 BKN010 BECMG 0608 SCT015 BKN020 PROB30 TEMPO 0816 1702SG40KT 4000 TSRA SCT010 BKN010CB BECMG 1821 3000 BR SKC=

Decode

Eighteen hour TAF issued at 2200 on 10th, London Heathrow Valid from oh six hundred to midnight the next day. Wind one three zero deg ten knots. Nine km. Broken at one thousand feet. Becoming from oh six hundred to oh eight hundred. Scattered at one thousand five hundred feet broken at 2000 ft. Prob thirty, temporary oh eight hundred to oh sixteen hundred, wind one seven zero deg twenty five knots max forty knots. Four thousand metres. Thunderstorm with rain. Scattered at one thousand feet. Broken cumulonimbus at one thousand five hundred feet. Becoming from eighteen hundred to twenty one hundred. Three thousand metres. Mist. Sky clear.

9HR TAF

FCUK33 EGGY 300900
EGGW 301019 23010KT 9999 SCT010 BKN018 6000 – RA BKN012 TEMPO 1418 2000 DZ OVC004 FM1800 3002OG30KT 9999 – SHRA BKN010CB=

Decode

Nine hour TAF issued at 0900 on the 30th. Luton Valid from ten hundred to nineteen hundred zulu on 30th. Wind two three zero deg ten knots. Ten km or more. Scattered at one thousand feet. Broken at one thousand eight hundred feet. Becoming from eleven hundred to fourteen hundred feet. Six km. Light rain. Broken at one thousand two hundred feet. Temporarily fourteen hundred to eighteen hundred. Two thousand metres. Moderate Drizzle. Overcast four hundred feet. From eighteen hundred. Three zero zero deg twenty kt max thirty kt. Ten km or more. Light rain showers. Broken CB one thousand five hundred feet.

The Air Pilot's **Manual**

Volume 2

Exercises and Answers

About the Exercises

These exercises form a vital part of the course. We suggest that you take a blank piece of paper and jot your answer down, with the number of the question beside it. This leaves your textbook unmarked and suitable for later revisions. The answers are at the end of each section (see previous page for page numbers).

Some questions involve multiple-choice answers. While these are not a good learning aid (continually reading incorrect statements is confusing), many of the exams you will do in the course of obtaining your licence and ratings will use this method of questioning. Here is an example:

1 In the signals area of an aerodrome, a white 'T' means:

   (a) landing direction is parallel with the shaft towards the cross-arm.

   (b) land on hard surfaces only.

   (c) land and taxi on hard surfaces only.

   (d) do not land.

A good technique in answering multi-choice questions is, prior to reading through the selection, think in your own mind what the answer might be. Then read the four choices, and quite often you will find the answer you already have in mind is amongst them. If not, then proceed to eliminate the incorrect statements. In the example above, the correct statement is **(a)**, so just record '**1. (a)**'.

Many other exercises contain **alternative answers**. These are shown in brackets, with an oblique stroke dividing the choices. Here is an example:

2 In general, an aeroplane (may/should not) land on a runway that is not clear of other aircraft.

Having read Chapter 2 on *Rules of the Air,* you already know the answer is '**should not**'.

Where an exercise requires you to think of the missing word(s) or numeral(s) in a statement, there is a series of dots to indicate where the missing item(s) go. For example:

3 The left wing navigation light is coloured ... .

NOTE Exercises containing a number of parts are shown as (i), (ii), (iii) etc., to distinguish them from multi-choice questions.

Prior to sitting for the examination in the particular subject, you should be achieving almost total success in these exercises.

Good luck!

Aviation Law, Flight Rules and Procedures

Exercises 1
Aviation Law and Legislation

1 The Chicago Convention was signed in ..... This led to the formation of ICAO, which is based in ....

2 The principles of the ICAO as laid down by the Chicago Convention are known as ....

3 Definitive documents relating to the ICAO Articles are known as ....

4 Each nation (State) that has signed the Chicago Convention (has/has not) exclusive sovereignty over the airspace above its territory.

5 A State's sovereignty over its territory (is/is not) considered the waters adjacent to its land areas.

6 Other than scheduled international flights, a member State (will/will not) allow aircraft from another State to fly into or through its airspace and on to land without prior permission.

7 Irrespective of its nationality an aircraft on an international flight (should/must) obey the regulations of the overflown State.

8 An aircraft operating within a State's territory must follow the rules of the air. Over the high seas these rules (do/do not) apply.

9 Each State (does/does not) have the right to search aircraft from other States on landing or prior to departure.

10 Each State (may/may not) require aircraft entering its territory to land at a customs airport for a customs examination unless the flight has permission to cross its territory without landing.

11 Aircraft (do/do not) have the nationality of the State in which they are registered.

12 Aircraft (may/may not) be registered in more than one State.

13 Aircraft operating internationally (should/must) display appropriate nationality and registration marks.

14 An aircraft landing in another State's territory (is/is not) normally exempt from customs duty on fuel, oil, spare parts and stores that were on board on arrival and are retained on board on departure.

15 Spare parts imported into a State for use by an aircraft from another State on international operations (will/will not) be free from duty.

16 Should an aircraft registered in one State be involved in an accident in another State, the State in which the accident occurs (shall/shall not) run an inquiry.

17 In the situation described in Question 16, the State in which the aircraft is registered (shall/shall not) be allowed to observe the inquiry.

18 All aircraft flying internationally (must/should) carry appropriate licences for each flight crew member, together with other aircraft documents such as the Certificate of Airworthiness and Certificate of Registration.

19 All aircraft operating internationally (shall/shall not) be provided with a Certificate of Airworthiness by the State in which the aircraft is registered.

20 Pilots and flight crew members engaged in international operations (must/should) hold licences issued by the State in which the aircraft is registered.

21 Certificates of Airworthiness and flight crew licences issued by the State in which the aircraft is registered (shall/shall not) be recognized by other contracting States, provided the requirements for the issue of such certificates and licences meet ICAO agreed standards.

22 All aircraft on international operations (shall/shall not) keep a journey logbook.

23 Weapons or munitions of war (may/may not) be carried in or above a State's territory except by permission of the State.

24 States (may/may not) prohibit or regulate the use of photographic apparatus in aircraft over their territory.

25 Aircraft that have failed to meet any international standard of airworthiness or performance at certification (shall/shall not) show on its airworthiness certificate full details of such failure(s).

26 Aircraft and flight crew (may/may not) operate internationally only if their certificates or licences permit it.

27 What do the letters ICAO stand for?

28 Official information regarding *The Rules of the Air and Air Traffic Services* is found in the ... section of the UK ....

29 NOTAM stands for .... to ......

30 Non-urgent information is often distributed to pilots as an .... Circular, and those directly associated with *air safety* are printed on .... paper.

Questions 31–45 are based on AICs, which you should read thoroughly

31 A pilot should not fly for at least .... hours after taking small amounts of alcohol, and proportionally longer if larger amounts are consumed.

32 A light aeroplane taking off behind a heavy aeroplane, but departing from an intermediate part of the same runway should allow .... minutes for the dispersal of wake turbulence.

33 A light aeroplane taking off behind a heavy aeroplane and using the full length of the same runway should allow .... minutes for the dispersal of wake turbulence.

34 A light aircraft following a heavy aircraft on final approach to land should stay well behind by a distance of at least .... nm and a time of .... min.

35 To avoid possible wake turbulence from a departing large aircraft, a pilot intending to land on the same runway should land (well beyond/at/well before) its rotation point.

36 Can a pilot fly if taking tranquillizers?

37 If a pilot wears full-lens spectacles for reading only, are they likely to be suitable for flying?

38 Is a small amount of frost, ice, snow or any other contamination on the upper leading edge of a wing dangerous?

39 A 10% increase in aircraft weight will increase the *take-off distance to 50 ft* by approximately ....%, and the landing distance from 50 ft by ...%.

40 A tailwind component equal to 10% of the lift-off speed will increase the take-off distance to 50 ft by ....%.

41 Soft ground or snow may increase take-off distance to 50 ft by .... % or more.

42 Avgas filler points should be painted ... to avoid confusion with turbine fuel filler points.

43 Avtur filler points should be painted ..... to avoid confusion with gasoline filler points.

44 Is Jet A-1 a turbine fuel or gasoline?

45 Take-off, climb and landing performance of light aeroplanes is discussed in
AIC …

46 ANO stands for … … … .

47 The ANO is sub-divided into P....
which consist of A....... numbered
consecutively from the beginning of the
ANO.

Exercises 2

Rules of the Air

1 The UK Rules of the Air apply to all
UK-registered aircraft:
  (a) wherever they may be.
  (b) only when in UK airspace.
  (c) only when in UK airspace and near
  offshore installations.

2 The UK Rules of the Air apply to (all/
only UK-registered) aircraft in UK
airspace and in the neighbourhood of
offshore installations.

3 Collision avoidance between two aircraft
is the responsibility of … ....

4 If two aircraft are approaching head-on,
each must turn …… .

5 When on converging courses, the 'give
way to the ….. rule' applies.

6 If two aircraft are well separated but on
converging courses, the aircraft with
right of way (should/need not) maintain
course and speed.

7 An aircraft which is obliged to give way
to another aircraft must avoid passing ....
or ….. or … ..…., unless passing well
clear of it.

8 Another aircraft is approaching at the
same level with a constant relative bearing of 30° right of the nose. A collision
risk (exists/does not exist).

9 A flying machine (must/need not) give
way to a glider and (must/need not) give
way to another flying machine towing a
glider.

10 A glider (has/does not have) right of way
over a flying machine and (has/does not
have) right of way over a balloon.

11 An aeroplane being overtaken (has/does
not have) right of way.

12 When overtaking another aeroplane
flying in the same direction at the same
level, keep clear by turning …… .

13 When overtaking another aeroplane
flying in the same direction at the same
level, keep the other aeroplane on the
….. .

14 An overtaking situation is said to occur
when the overtaking aircraft is within ....
degrees of the overtaken aircraft's
centreline.

15 If two aircraft in flight are well separated
but on a collision course, the aircraft
with the other on its left should:
  (a) give way by turning right.
  (b) give way by turning left.
  (c) maintain its course and speed.
  (d) climb.

16 If two aircraft in flight are well separated
but on a collision course, the aircraft
with the other on its right should:
  (a) give way by turning right.
  (b) give way by turning left.
  (c) maintain its course and speed.
  (d) climb.

17 When flying in the vicinity of an aerodrome a pilot (should/need not)
conform with the traffic pattern, otherwise keep well clear.

18 The turns in most aerodrome traffic
patterns are to the (left/right).

19 An arriving aircraft approved by ATC to
make a straight-in approach to a runway
will normally report "long final" at
.... nm, and "final" at .... nm.

20 A landing aircraft (has/does not have)
right of way over an aeroplane taxiing
for take-off.

21 If two aeroplanes are on final approach together, the (lower/higher) one has right of way, unless otherwise instructed by ATC or in an emergency.

22 In general, an aeroplane (may/should not) land on a runway that is not clear of other aircraft.

23 When approaching to land on an aerodrome at which take-off and landing operations are not confined to a runway (i.e. it is an 'all-over field'), you should land to the (left/right) of another aircraft that has just landed.

24 When approaching to land on an all-over field, you should keep another aircraft that has just landed on your (left/right).

25 After landing on an all-over field, an aeroplane should turn (left/right).

26 When following a line feature, keep to the (right/left) of it.

27 When following a line feature, keep it on your (left/right).

28 The left wing navigation light is coloured ..... .

29 The right wing navigation light is coloured ....

30 The red navigation light is visible from straight ahead through an arc of ....°.

31 The tail navigation light is coloured ..... and is visible from behind and in an arc ....° either side.

32 Night is defined (for the Rules of the Air) as .... minutes after sunset until .... minutes before sunrise.

33 At night you see the red navigation light of an aeroplane whose range is decreasing out to your left. A risk of collision (does/does not) exist.

34 At night you see the red navigation light of an aeroplane whose range is decreasing out to your right. A risk of collision (does/does not) exist.

35 At night you see the green navigation light of an aeroplane whose range is decreasing on a relative bearing of 030° (i.e. 30° to the right of your nose). A risk of collision (does/does not) exist.

36 At night you see two navigation lights of an aircraft whose range is decreasing, the green on your left and the red on your right. The situation is:

(a) a risk of collision exists and you should turn right.

(b) a risk of collision exists and you should immediately turn left.

(c) no risk of collision exists.

(d) you should turn right and overtake the aeroplane ahead.

37 At night you see the white navigation lights of an aircraft about 3 nm ahead whose range is decreasing. The situation is:

(a) a risk of collision with the aeroplane ahead exists and you should turn right to overtake it.

(b) a risk of collision exists and you should immediately turn left.

(c) no risk of collision exists.

38 If a navigation light fails in flight at night, you should advise ATC. If for some reason radio contact is not made, then you should:

(a) continue with the flight as planned.

(b) land as soon as possible at a suitable aerodrome.

(c) fire a distress flare.

(d) switch the other navigation lights OFF.

39 Free balloons (may/are required to) show a steady red light in all directions at night.

40 Gliders (may/are required to) show a steady red light in all directions at night.

41 Out to your left at night you see an anti-collision light plus the green and white navigation lights of another aircraft. It is an (aeroplane/airship/glider). Which of you has right of way?

42 An aircraft must not fly closer than .... ft to any person, vessel, vehicle or structure.

43 An aeroplane flying over a congested area such as a town (should/need not) fly at a height that would enable it to land clear if the engine failed.

44 As well as flying at a height over a congested area that would allow it to land clear in the event of an engine failure, an aeroplane should not normally fly lower than ..... feet above the highest fixed object within .... metres of it. An exception is when taking off or landing at a (government/CAA/licensed/unlicensed/private) aerodrome.

45 No aircraft should normally fly over or within (3,000/2,000/1,000) metres of an open-air gathering of more than (3,000/2,000/1,000) people.

46 Aerobatics are not permitted (below 5,000 ft amsl/in controlled airspace/over a congested area such as a city, town or settlement).

47 Aerobatics within controlled airspace (must/need not) have specific approval from the controlling authority.

48 In general, the minimum height over a congested area is ..... feet above the highest fixed object within .... metres.

49 In general, the minimum height over a large open-air gathering is .... metres.

50 The responsibility for an aeroplane maintaining sufficient height to glide clear of a congested area in the event of an engine failure lies with:

(a) the pilot-in-command.

(b) Air Traffic Control.

51 When simulated instrument flying is taking place, the aeroplane (must/need not) have dual flying controls and, in the second control seat, a (safety pilot/unqualified observer) must be seated.

52 When practising instrument flying under simulated instrument flight conditions, with the pilot who is flying having his view outside of the cockpit restricted, a safety pilot should be in the second control seat. An observer (may/must not) be carried to improve the lookout.

53 In order to comply with ICAO standards, pilots leaving UK airspace to make international flights (are/are not) required to carry a copy of the procedures to be followed in the event of interception of their aircraft for reasons of military necessity.

54 If you are intercepted by another aircraft you (must/may) immediately notify, if possible, the appropriate Air Traffic Services Unit.

55 Visual signals (may/may not) be used to warn unauthorized aircraft flying in, or about to enter, a restricted, prohibited or danger area. These signals are a series of projectiles discharged at intervals of .... seconds, each showing, on bursting, ... and ..... stars or lights.

56 The UK Air Navigation Order (ANO) defines night as the time between .... minutes after sunset and .... minutes before sunrise. However, the ICAO definition is the time between the ... of evening civil twilight and the .... of morning civil twilight, or as prescribed by the appropriate aviation authority.

Exercises 3

Aerodromes

1 Is prior permission required to use a military aerodrome under normal operating conditions?

2 Is prior permission required to use an unlicensed aerodrome under normal operating conditions?

3 Is permission required to use a civil aerodrome with an ordinary licence under normal operating conditions?

4 A civil aircraft may land at an aerodrome not listed in the UK AIP in an e........ in flight or by obtaining p.... p......... from the aerodrome operator.

5 From which of the following types of aerodrome may aeroplanes or rotorcraft being used for flight instruction operate: government aerodrome, CAA aerodrome, private landing ground not licensed for training, an aerodrome licensed for training?

6 Is a runway considered to be part of the manoeuvring area of an aerodrome?

7 An aerodrome sign composed of white lettering on a red background is a .... instruction sign.

8 A sign consisting of a black letter 'C' on a square (yellow/red) background denotes 'control tower' or other aerodrome administrative building to which pilots of visiting aircraft should report.

9 Is a taxiway considered to be part of the manoeuvring area of an aerodrome?

10 On a taxiway, what does two solid yellow lines followed by two broken yellow lines indicate?

11 Is an apron or maintenance area considered to be part of the manoeuvring area of an aerodrome?

12 Identification beacons at military aerodromes flash a (one/two/three) morse code identification group every 12 seconds in the colour ... .

13 Identification beacons, when in use at civil aerodromes, flash a (one/two/three) morse code identification group every 12 seconds in the colour ....

14 Aerodrome beacons give an alternating flash signal using the colours ..... and ......, or white and white.

15 When two aircraft are approaching head-on or nearly so when taxiing, each must turn ....

16 Overtake when taxiing by turning to the .... and passing so that the other aircraft is on your ....

17 When taxiing, give way to the ....

18 A landing aeroplane has right of way over a taxiing aeroplane. (True/False)?

19 Which has right of way – a taxiing aeroplane or an aeroplane under tow by a tractor.

20 What is the priority for right of way on the ground for the following: vehicles, vehicles towing aircraft, aircraft taking off or landing, aircraft taxiing?

21 The dimensions of an ATZ surrounding an aerodrome whose longest runway is 1,850 metres or less are ..... ft aal high out to a circle of radius .... nm.

22 The dimensions of an ATZ surrounding an aerodrome whose longest runway is greater than 1,850 metres are ..... ft aal high out to a circle of radius .... nm.

23 In the signals area of an aerodrome, a white 'T' means:

   (a) landing direction is parallel with the shaft towards the cross-arm.

   (b) land on hard surfaces only.

   (c) land and taxi on hard surfaces only.

   (d) do not land.

24 In the signals area of an aerodrome, a white dumb-bell means:
(a) landing direction is parallel with the shaft towards the cross-arm.
(b) land on hard surfaces only.
(c) land and taxi on hard surfaces only.
(d) do not land.

25 In the signals area of an aerodrome, a red square with a single yellow diagonal stripe means:
(a) do not land.
(b) take special care when landing because of the poor state of the manoeuvring area.
(c) gliders are operating.
(d) helicopters are operating.

26 The addition of black stripes to each circular portion of a white dumb-bell at right angles to the shaft means that aeroplanes should take off or land on (a runway/any surface be it hard or soft), and that movement on the ground (is/is not) confined to hard surfaces.

27 In the signals area of an aerodrome, a red square with a diagonal yellow cross means:
(a) do not land.
(b) take special care when landing because of the poor state of the manoeuvring area.
(c) gliders are operating.
(d) helicopters are operating.

28 A red and yellow striped arrow bent through 90° around the edge of the signals area and pointing in a clockwise direction means that:
(a) landing direction is parallel with the shaft towards the cross-arm.
(b) a large obstacle is out to the right.
(c) land and taxi on hard surfaces only.
(d) right circuits are in force.

29 A black ball suspended from a mast in the signals area of an aerodrome means ...

30 Two or more white crosses along a runway, taxiway, or section thereof, with the arms at an angle of 45° to the centreline as shown below, signifies ...

31 A rectangular red/yellow chequered flag or board in the signals area of an aerodrome means ...

32 An aerodrome marked as 'PPR' means that:
(a) private pilots only should use the aerodrome.
(b) prior permission is required to use the aerodrome.
(c) its identifying code is 'PPR'.
(d) it is unserviceable.

33 Aerodromes not listed in the AD section of the UK Aeronautical Information Publication may be used:
(a) at any time.
(b) only if prior permission has been obtained or in an emergency.
(c) only in an emergency.
(d) never.

34 Is an apron or maintenance area considered to be part of the manoeuvring area of an aerodrome?

35 Permission (should/need not) be obtained from ATC to fly in an Aerodrome Traffic Zone.

36 An aeroplane is flying at 3,000 ft above mean sea level directly overhead an aerodrome which has an elevation of 1,263 ft amsl. Is the aeroplane within the Aerodrome Traffic Zone?

37 A long white strip with two short strips across it, i.e. a double white cross, means …

38 An unfit section of taxiway or runway is indicated by two ..... crosses.

39 A square yellow board with a black 'C' indicates the building where a pilot can report to ATC or other aerodrome authority. (True/False)?

40 When dropping a tow rope at an aerodrome the pilot (should/need not) fly in a direction appropriate for landing and drop the tow rope in the area designated by a (yellow/red/white/yellow and red) cross, or as directed by ATC or the person in charge.

41 At an aerodrome where the circuit is variable, a … flag indicates that a left-hand circuit is in operation, and a ..... flag indicates that a right hand circuit is in force.

42 A continuous red light directed from the Tower to an aeroplane taxiing means ..... .

43 A flashing green light directed from the Tower to an aeroplane taxiing means …

44 A flashing green light directed from the Tower to an aeroplane in flight means …

45 A continuous green light directed from the Tower to an aeroplane in flight means …

46 Red flashes directed to an aeroplane in flight mean …

47 A continuous red light directed to an aeroplane in flight means …

48 A red flare or red pyrotechnic light directed to an aeroplane in flight means …

49 A flashing white light to an aircraft on the ground from the Tower means …

50 A flashing white light to an aircraft in flight from the Tower means …

51 To indicate that he is compelled to land a pilot should:
(a) switch his landing and/or navigation lights ON and OFF.
(b) take no special action.

52 To indicate that he is compelled to land and is in need of immediate assistance a pilot should:
(a) switch his landing and/or navigation lights ON and OFF.
(b) take no special action.
(c) fire a red flare.

53 The marshaller's hands held vertically above his head means …

54 The marshalling signal of both arms repeatedly crossed above the head means:
(a) turn right.
(b) turn left.
(c) move ahead.
(d) stop.

55 The marshalling signal of both arms repeatedly and rapidly crossed above the head means …

56 The marshalling signal of both arms moving upward and back means:
(a) turn right.
(b) turn left.
(c) move ahead.
(d) stop.

57 The marshaller's right arm down (i.e. on your left) and his left arm repeatedly moving upward and backward means:

(a) turn left.

(b) turn right.

(c) stop.

(d) increase power.

58 The marshaller's arms placed down towards the ground palms down, and then moved slightly up and down several times means:

(a) stop.

(b) stop the engine.

(c) slow down.

(d) increase speed.

59 To signal a marshaller to remove chocks, you would …

60 To signal a marshaller to insert chocks, you would …

61 Flying training for the purpose of obtaining a licence may occur at which of the following:

(a) Aerodromes licensed for the purpose by the CAA.

(b) A government aerodrome notified for the purpose.

(c) A CAA-operated aerodrome notified for the purpose.

(d) Any landing area.

62 Avgas fuelling equipment is usually marked … .

Exercises 4

Altimeter-Setting Procedures

1 As height is gained, atmospheric pressure (increases/decreases/stays the same).

2 Atmospheric pressure at a given place (changes/does not change) from time to time.

3 The altimeter subscale (is/is not) used to set the pressure datum from which the height will be measured.

4 To measure the vertical distance above mean sea level (amsl) the altimeter subscale should be set to (QNH/QFE/1013).

5 An altimeter reading based on QNH is called (altitude/height/elevation).

6 An altimeter reading based on QFE is called (altitude/height/elevation).

7 The vertical distance of an aerodrome above sea level is called its … .

8 To measure the vertical distance above aerodrome level (aal) the altimeter subscale should be set to (QNH/QFE/1013).

9 To measure the *pressure altitude* when flying at a flight level, the altimeter subscale should be set to (QNH/QFE/1013).

10 A pressure altitude of 3,500 ft is flight level ….. .

11 A pressure altitude of 24,500 ft is FL….

12 The altitude below which aeroplanes cruise with QNH set in the altimeter subscale is called the … … .

13 The level above which aeroplanes cruise with 1013 set in the altimeter subscale is called the … … .

14 Choose the cruising level carefully to ensure that there is adequate separation from … and other … .

15 For take-off and landing the altimeter subscale may be set to either the aerodrome … or the aerodrome …. pressure settings.

16 When cruising below 3,000 ft amsl the altimeter subscale should be set to the … … .

17 If the QNH is not periodically adjusted as the aeroplane flies towards an area of lower pressure, the aeroplane will:

(a) gradually climb.

(b) gradually descend.

(c) maintain the same height amsl.

18 Flying beneath a Terminal Control Area (TMA) the pilot should set:
(a) regional QNH.
(b) aerodrome QNH.
(c) 1013.
(d) aerodrome QFE.

19 Penetrating a Military Aerodrome Traffic Zone (MATZ) the pilot will generally be asked to set:
(a) regional QNH.
(b) aerodrome QNH.
(c) 1013.
(d) aerodrome QFE.

20 Explain what is meant by *clutch* QFE.

21 With aerodrome QNH set, the altimeter should read:
(a) zero when the aeroplane is on the ground.
(b) aerodrome elevation when the aeroplane is parked.
(c) aerodrome elevation when the aeroplane is parked at the highest point on the landing area.
(d) zero when the aeroplane is parked at the highest point on the landing area.

22 With aerodrome QFE set the altimeter should read:
(a) zero when the aeroplane is on the ground.
(b) aerodrome elevation when the aeroplane is parked.
(c) aerodrome elevation when the aeroplane is parked at the highest point on the landing area.
(d) zero when the aeroplane is parked at the highest point on the landing area.

23 The quadrantal rule for aeroplanes flying under the Instrument Flight Rules and cruising above the transition level is based on (magnetic/true) track.

24 An IFR aeroplane on a track of 040°M to satisfy the quadrantal rule should cruise at:
(a) FL50.
(b) FL55.
(c) FL60.
(d) FL65.

25 An IFR aeroplane on a track of 090°M to satisfy the quadrantal rule should cruise at:
(a) FL50.
(b) FL55.
(c) FL60.
(d) FL65.

26 An IFR aeroplane flying with a heading of 085°M in a strong wind blowing from the north experiences 8° right drift. To satisfy the quadrantal rule it should cruise at:
(a) FL50.
(b) FL55.
(c) FL60.
(d) FL65.

27 An aeroplane has a true track of 265°T in an area where the magnetic variation is 7°W and therefore is a magnetic track of (265 + 7) = 272°M. To satisfy the quadrantal rule, it should cruise at:
(a) FL35.
(b) FL40.
(c) FL45.
(d) FL50.

28 Convert FL50 to an altitude if the QNH is 980 mb(hPa).

29 Convert FL50 to an altitude if the QNH is 1033 mb(hPa).

30 When an aircraft is descending to below the transition level, before commencing approach for a landing at a civil aerodrome, ATC will pass (aerodrome QNH/aerodrome QFE/regional QNH).

31 In the UK it is usual to make a visual approach using an altimeter setting of either aerodrome ... to indicate height above the runway, or aerodrome ... to indicate height above mean sea level.

Exercises 5

Airspace

1 The basic division of airspace in the UK is into the ...... and .... Flight Information Regions, abbreviated ...s.

2 FIRs in the UK extend upwards to a limit of ...... feet, flight level ..... .

3 The airspace above the FIRs is known as .... , and abbreviated as ... which stands for .... Region.

4 Airspace is categorised as either un.......... or c.......... .

5 The ICAO airspace-classification system grades airspace from A to G. Controlled airspace is allocated from Class .... to Class ...., and uncontrolled airspace is Class .... and Class ..... .

6 Upper airspace, i.e. above FL245, is Class .... and is (controlled/uncontrolled).

7 ICAO Class C (is/is not) currently allocated in the UK.

8 The busiest airspace commonly used by airliners, such as Airways and busy TMAs and Control Zones, are Class .... airspace, and are (controlled/uncontrolled). This airspace (is/is not) available to VFR aircraft.

9 Less busy controlled airspace is allocated Class .... or Class ..... .

10 Uncontrolled airspace is Class .... or Class ...., with Advisory Routes being allocated Class ...., and Open-FIR being known as Class ..... .

11 A Flight Information Region (FIR) (will/will not) provide flight information, advisory information and an alerting service (search and rescue) for known aircraft.

12 A well-used route in uncontrolled airspace, between specified locations, in which a traffic separation service is available to all participating aircraft is called an ...., abbreviated .... . Such a route is deemed to be .... nm wide.

13 Airspace that is neither controlled airspace nor advisory airspace is Class .... and is referred to as the ....-.... .

14 The term ATZ is an abbreviation for ....

15 Describe the dimensions of an ATZ.

16 An aircraft fitted with radio, entering or leaving an ATZ, must transmit .... and ...... to the appropriate Air Traffic Service Unit.

17 The term MATZ is an abbreviation for ....

18 Describe the dimensions of a MATZ with one stub.

19 A MATZ (does/does not) contain an ATZ.

20 When transiting a MATZ, a pilot will be requested to set .... in the altimeter subscale.

21 An area around certain aerodromes from ground level to a specified altitude within which an Air Traffic Control service is provided to all flights is called:
(a) a Control Zone.
(b) a Terminal Control Area.
(c) an Aerodrome Traffic Zone.

22 The minimum lateral dimension of a Control Zone (CTZ) is (2/5/10) nautical miles either side of the centre of the aerodrome in the direction of the approach path.

23 A Control Zone is abbreviated as ... and extends from .... to a specified altitude or flight level.

24 A Control Area is abbreviated as ... and extends upwards from a specified .... or ....

25 A Control Area in the form of a corridor between major aerodromes is called an ....

26 A Control Area established at the confluence of Airways in the vicinity of major aerodromes is called a .... and is commonly abbreviated to ... .

27 The minimum radio equipment required to transit Class D airspace below FL100 is ....

28 The dimensions of an Airway are .... nm either side of a straight line joining certain places and (has/does not have) specified vertical limits.

29 As an Airway approaches a TMA, its lower level (may/will not) descend.

30 A private pilot without an Instrument Rating (may/may not) enter an Airway.

31 A Special VFR Clearance will only be issued for flight into certain:
(a) Airways.
(b) CTRs.
(c) TMAs.
(d) CTAs.

32 The holder of a basic PPL (may/may not) fly at right-angles across the base of an Airway, the lower limit of which is specified as a flight level (FL).

33 A pilot with an Instrument Rating, even in an aeroplane not fully equipped for IFR flight, (may/may not) cross an Airway by penetrating it, provided the conditions are VMC by day, he has filed a flight plan and obtained an ATC crossing clearance.

34 An area in which flight is prohibited is called a .... Area.

35 An area in which flight is restricted according to certain conditions is called a .... Area.

36 Airspace in which activities dangerous to flight may occur within scheduled hours are known as ...... Areas and are shown on some aeronautical charts with a solid ... outline.

37 Airspace in which activities dangerous to flight may occur as advised by NOTAM are known as ...... Areas and are shown on some aeronautical charts with a .... outline.

38 A section of airspace designated as D703/15 is a ...... Area located between latitudes .... and .... extending from the surface to an altitude of ..... ft amsl.

39 An airspace designated as D505A/1–4.5 is a ...... Area located between latitudes .... and .... extending from .... to ..... .

40 Known permanent obstructions .... (agl/amsl) or higher are published in the UK AIP and shown on aeronautical charts.

41 Permanent obstructions over .... are lit.

42 Land-based air navigation obstacles are defined as any obstacle .... ft agl or higher, with permanent obstacles above .... ft agl being lit.

43 Known hang-gliding sites are depicted on aeronautical charts. Thermalling hang-gliders may be at heights up to (the cloud base or the base of controlled airspace, whichever is the lower/the cloud base or 5,000 ft, whichever is the lower/5,000 ft).

44 Parascenders may be at heights up to ...

45 Launch cables for parascenders may be carried up to .... ft (amsl/agl).

46 Winch-launched gliders may carry the cable up to a height of .... ft (agl/amsl) or more before releasing it.

47 Pilots should avoid flying at less than ..... ft agl over areas where birds are known (or likely) to concentrate.

48 Military low flying occurs in many parts of the UK at heights up to .... ft agl, but with the greatest concentration between .... and .... ft agl, a height band that civil pilots are strongly recommended to avoid when possible.

49 Civil pilots (are/are not) permitted in AIAAs, which are A.... of I...... A..... A........ .

50 Given a choice of two routes, one passing through an AIAA and the other not, it would be good airmanship to choose the (first/second) route.

51 If you do fly through an AIAA, you (should/need not) maintain an especially good lookout, and (should/should not) make use of any radar service available.

52 Aircraft towing targets for military practice may trail up to ..... ft of cable and fly at up to ...... ft amsl. The areas in which target-towing trials may occur (are/are not) listed in the UK AIP.

53 Royal Flights are conducted, where possible, within existing controlled airspace but, if this is not possible, airspace known as .... is established along the route. Details (are/are not) promulgated by NOTAM.

54 Free-fall parachuting may occur at heights up to ....

Exercises 6

Air Traffic Services

1 The objectives of air traffic services shall be to:

(i) prevent .... between aircraft in the air and on the ground;

(ii) prevent .... between aircraft and objects on the .... area of aerodromes;

(iii) expedite and .... an orderly flow of traffic;

(iv) provide advice and information useful for the safe and efficient conduct of ....;

(v) notify and cooperate with appropriate .... regarding aircraft in need of ...... and rescue.

2 ATS airspaces shall be classified. Complete the following.

(a) **Class A.** ... flights only permitted; all flights are subject to ATC service and are separated from each other.

(b) **Class B.** ... and ... flights; all flights are subject to ATC service and are separated from each other.

(c) **Class C.** IFR and VFR flights; all flights subject to ... service. IFR flights are separated from other IFR flights and from VFR flights. VFR flights are separated from IFR flights and receive traffic information about other VFR flights.

(d) **Class D.** IFR and VFR flights; all flights subject to ATC service. IFR flights are separated from other ... flights and receive traffic information in respect of ... flights. VFR flights receive traffic information about all other flights.

(e) **Class E.** IFR and VFR flights; all flights subject to ... service. IFR flights are separated from other IFR flights. All flights receive traffic information as ....

(f) **Class F.** IFR and VFR flights; all participating IFR flights receive an .... advisory service and all flights receive flight information service .. ..

(g) **Class G.** ... and ... flights; all flights receive ...... information service .. ..

3 ATC is the abbreviation for ....

4 ATC at an aerodrome may be shared between .... and ....

5 Information is usually passed from ATC to aircraft in flight by means of radio, but, for non-radio flights or in the case of radio failure, may also be passed by .... or .... signals from the Control Tower and by .... signals in the aerodrome signals area.

6 FIS is the abbreviation for ....

7 ATC provides control. FIS (does/does not) provide control.

8 A/G radio (does/does not) provide control at an aerodrome.

9 'Southampton Approach' refers to (ATC/FIS/A/G).

10 'Binbrook Tower' refers to (ATC/FIS/A/G).

11 'Newtownards Radio' refers to (ATC/FIS/A/G).

12 'Rochester Information' refers to (ATC/FIS/A/G).

13 An air/ground radio station, which is operated at an aerodrome by personnel without ATC qualifications and who provide neither ATC nor FIS, is indicated by them using the callsign of the place name followed by (tower/approach/radar/radio/ground).

14 A VFR pilot receiving Radar Advisory Service (may/should not) accept a radar controller's instruction that would take him into cloud.

15 A pilot in receipt of a Radar Information Service (is/is not) responsible for collision avoidance.

16 Who is responsible for the separation of conflicting traffic if one aircraft is in receipt of a Radar Information Service and the other is not: (radar controller/pilot-1/pilot-2/both pilots)?

17 For which of the following is a VFR pilot responsible, if in receipt of a Radar Advisory Service outside controlled airspace: (terrain clearance/maintaining the required visibility/obtaining permission to enter Class D controlled airspace or an ATZ)?

18 When departing on a flight, a pilot will normally '.... out' with ATC.

19 A pilot is *advised* to submit a flight plan if intending to fly over sparsely populated or mountainous areas or more than .... nm from the coast.

20 A flight plan (must/need not) be submitted for a night training flight.

21 A flight plan (must/may/need not) be submitted for a flight during which it is intended to use the Air Traffic Advisory Service on an Advisory Route.

22 A flight plan (must/may/need not) be submitted for a flight to or from the UK which will cross a UK FIR boundary.

23 A flight plan (must/may/should not/need not but should) be submitted for a flight more than ten miles from the coast or over sparsely populated or mountainous areas.

24 Flight plans should normally be submitted to ATC at least .... minutes before requesting a taxi or start-up clearance.

25 Due to poor weather or some other cause, a flight for which a flight plan has been submitted diverts and lands at an aerodrome other than the destination specified. The ATSU at the planned destination must be told within .... minutes of the Estimated Time of Arrival there.

26 The primary method of private pilots obtaining a preflight meteorological briefing is by (self-briefing/personal briefing by a meteorologist); this (does/does not) require prior notice.

27 A request for a special forecast for a flight of 623 nm, proceeding outside the area of coverage of UK area forecasts, should be made at least .... hours prior to flight.

28 A request for a special forecast for a flight of 315 nm, proceeding outside the area of coverage of UK area forecasts, should be made at least .... hours prior to flight.

29 The *AIRMET* weather forecast service (is/is not) available via the public telephone network.

30 If you experience severe weather conditions en route you should:
(a) advise the appropriate ATC unit by radio as soon as possible.
(b) fill in a Met Form 215 immediately after you land.

31 Simplification of formalities for international air transport is called F........... .

Exercises 7

Visual Flight Rules (Rules 24–27)

1 Conditions in which flight is possible under VFR, the Visual Flight Rules, are known as V..... M............. C.......... .

2 Class A airspace (is/is not) available to VFR flights.

3 Most VFR flights in UK controlled airspace will occur in Class .... and Class .... airspace.

4 Flight in uncontrolled airspace will occur in Class .... and Class .... airspace.

5 For VFR flight above flight level 100, the VFR minima are:
– a flight visibility of .... km;
– a horizontal distance from cloud of ..... metres;
– a vertical distance from cloud of ..... feet.

6 Below FL100, the VFR minimum for flight visibility is .... km.

7 The VFR minima requirements are reduced further at or below 3,000 ft amsl for (uncontrolled/controlled/all) airspace.

8 The VFR minima requirements for flight at or below 3,000 ft in uncontrolled airspace, i.e. Class .... and Class .... airspace, are flight visibility of .... km, clear of ..... and in sight of ... .

9 For flight at or below 3,000 ft amsl in uncontrolled airspace at indicated airspeeds of 140 kt or less, the flight visibility requirement is ..... metres; it (may/may not) be less for helicopters.

10 A VFR pilot needs to take account of (VFR minima/weather minima privileges of his licence/both).

11 VFR minima are specified in (Rules of the Air/Air Navigation Order – Schedule 8).

12 The privileges of the Private Pilot's Licence, including minimum flight visibility, are specified in (Rules of the Air/ANO Schedule 8).

13 A PPL holder, with no IMC rating, flying at or below 3,000 ft amsl at an IAS of 140 kt or less in uncontrolled airspace, must satisfy the following conditions:
(i) flight visibility ... ;
(ii) distance from cloud ... ;
(iii) in sight of the surface (is/is not) a requirement.

14 Do the above requirements change if a passenger is carried?

15 VFR flight (may/may not) occur on Airways, which are Class .... airspace when not penetrating airspace of lesser classification.

16 SVFR stands for ....

17 To enter a Control Zone in weather conditions below VFR, the VFR pilot requires a .... Clearance.

18 An SVFR clearance through a Control Zone (is/is not) a concession granted by ATC.

19 SVFR flight (is/is not) permitted in an Airway.

20 An SVFR clearance at a low level (absolves/does not absolve) the pilot from the 1,500 ft low flying rule over congested areas provided he can still land clear in the event of an engine failure.

21 A flight plan (is/is not) required for an SVFR flight.

22 The responsibility for maintaining adequate flight visibility and distance from cloud on a Special VFR clearance rests with (ATC/the pilot).

23 The minimum flight visibility generally required for an SVFR flight in a control zone is:
(a)   10 km.
(b)   5 km.
(c)   1,800 m.
(d)   1,500 m.

24 The usual minimum flight visibility for an SVFR clearance may be relaxed at some aerodromes in control zones where there are entry/exit lanes speci-fied in the UK AIP as allowing flight in Instrument Meteorological Conditions without full IFR procedures being followed. (True/False)?

25 An aircraft on an SVFR clearance in a control zone (may/must not) fly lower than 1,500 ft over obstacles in a congested area, and (must/need not) be able to land clear of the congested area in the event of an engine failure.

26 What is the ANO definition of night?

27 There is no flight under the Visual Flight Rules (VFR) at night. (True/False)?

28 If sunset occurs at time 1800 UTC, then a VFR pilot must plan to be on the ground no later than .... UTC.

29 VFR pilots (are/are not) advised to select cruising levels in accordance with the quadrantal rule, in the interests of safety.

30 The most suitable cruising level for track 273°M, at or above FL40 is .... .

31 For a planned track of 085°T in an area where variation is 5°W, the most suitable cruising level at or above FL40 is .... .

Exercises 8

Instrument Flight Rules (Rules 28–32)

1 IFR stands for ....

2 In general, an IFR flight must not fly at less than ..... ft above the highest obstacle within .... nm.

3 The requirements in Question 2 for an IFR flight are reduced during take-off and landing, and when flying:
(i)   at ..... ft amsl or below; and
(ii)   (clear/not within 1,500 m) of cloud; and
(iii)   (within/not within) sight of the sur-face.

4 A pilot flying under the Instrument Flight Rules outside controlled airspace should have 1013 set on the altimeter subscale and fly at a flight level according to the quadrantal rule:
(a)   at all heights.
(b)   above 3,000 ft amsl.
(c)   above the transition altitude.
(d)   above 3,000 ft amsl or the transition altitude, whichever is the higher.
(e)   above 5,000 ft amsl.

5 IFR cruising level is based on (magnetic track/true track/magnetic heading/true heading).

6 When following the quadrantal rule, the pilot of an aeroplane on a magnetic track of 140°M should cruise at flight level (70/75/80/85) with (1013/QNH) set on the altimeter subscale.

7 What is a suitable cruising level near to FL60 if your planned track is 267°T in an area where magnetic variation is 6°W, and you are experiencing 10 degrees of port drift? What level near FL60 is suitable if the drift is only 2 degrees to port?

8 What is a suitable cruising level if the track is 178°T, in an area where magnetic variation is 5°W: FL(70/75/80/85)?

Exercises 9

Registration and Airworthiness

1 What shall an aircraft nationality and registration mark consist of?

2 On heavier-than-air aircraft the marks shall appear once on the ..... surface of the wing, on the .... side unless they extend across the whole wing structure.

3 On lighter-than-air aircraft, the height of the marks shall be at least .... centimetres.

4 What should the aircraft identification plate be made of and where should it be placed in the aircraft?

5 Are aeroplanes and gliders required to be registered?

6 A flying machine is considered to be any power-driven, heavier-than-air aircraft. (True/False)?

7 Notification in writing to the CAA regarding any change in ownership or part-ownership, or the destruction or permanent withdrawal from use of an aircraft (is/is not) required by the Air Navigation Order.

8 A Certificate of Airworthiness in the UK usually lasts for .... years.

9 Generally a Noise Certificate is required for all aircraft, with the exception of certain .... aeroplanes.

10 For an aeroplane to tow a glider, its CofA must be endorsed to this effect. (True/False)?

11 Unserviceable items should be recorded after flight in the ....

12 Defects should be entered in the Technical Log:
(a) before the aeroplane is flown by another pilot;
(b) at the end of the day;
(c) the earlier of (a) or (b).

13 If an aircraft is modified, repaired or maintained in any way that might affect its airworthiness, such as replacement of an aileron or installation of devices to reduce drag, then its Certificate of Airworthiness (will/will not) be suspended awaiting CAA approval and issuance of a Certificate of ....

14 A pilot may carry out only minor repairs and replacements as specified in the ANOs for aircraft with maximum total weight authorised of ..... kg or less and with a Certificate of Airworthiness in the .... or .... categories.

15 Is a pilot permitted to replace VHF communications equipment if it is not combined with navigation equipment?

16 Should the operator of an aeroplane maintain an Engine Logbook?

17 That part of the Certificate of Airworthiness which approves certain weight limitations is known as the ....
....

18 Operating outside the conditions specified in the CofA may, apart from making the pilot liable for imprisonment or a fine, invalidate ...., ... and .... on the aircraft and its equipment.

19 For how long should an aircraft, engine or propeller logbook be preserved after the particular aircraft, engine or propeller has been destroyed or permanently withdrawn from use.

20 In relation to light aircraft, whose responsibility is it to ensure that the aircraft's documentation is valid and that the aeroplane is fit to fly in every respect on a planned flight?

(a) the owner or operator

(b) the aircraft's commander

21 A Weight Schedule prepared by the operator of an aeroplane should be preserved until (0/6/12/18/36) months after the next official weighing.

22 If an aircraft is damaged, who determines whether such damage renders the aircraft no longer airworthy?

23 What do loading limitations include?

24 Airspeed limitations shall include all speeds that are limiting in relation to .... integrity and ...... qualities.

25 What does aircraft *empty mass* (weight) usually exclude?

Exercises 10

Pilots' Licences

1 The minimum age for the granting of a Private Pilot's Licence is .... years.

2 Assuming a current medical certificate is held, the minimum age at which solo flight is permitted is 16 years. (True/False)?

3 The period of validity of a JAR–FCL Private Pilot's Licence, provided the medical and flight test or experience requirements are met, is:

(a) 3 years.

(b) 5 years.

(c) life.

4 The privileges of a JAR–FCL Private Pilot's Licence may only be exercised if certain medical, flight test or recency requirements are met. A Certificate of Revalidation or a Certificate of Test remains valid for .... months.

5 The Certificate of Revalidation for a single pilot, single-engine piston rating may be renewed by:

completing .... hours of flight time which must include at least .... hours as pilot-in-command, ... take-offs and landings and a minimum 1-hour training flight with a flying instructor in the 12 months preceding the expiry date of the rating; **or**

passing a proficiency check with an authorised examiner within a period of .... months preceding the date of expiry of the rating.

6 To exercise the privileges of a valid JAR–FCL PPL, the holder must have at least two additional documents that are also valid; they are ...

7 The period of validity is .... after the date of the CofT or CofR.

8 A private pilot (without an IMC or Instrument Rating) may fly in cloud. (True/False)?

9 The privileges of the PPL are contained in Schedule 8 of the ... .

10 A JAR private pilot (without an IMC or Instrument Rating) when flying outside controlled airspace, at or below 3,000 ft amsl and at or below 140 knots indicated airspeed, may fly in a minimum flight visibility of .... km.

11 In the above case, what is the minimum flight visibility if a passenger is carried?

12 A private pilot (without an IMC or Instrument Rating) may only carry a passenger at night if he has a Night Rating and has within the last 90 days carried out at least …. take-offs and landings as the sole manipulator of the controls, of which …. take-off and …. landing must have been at night.

13 The ANO defines night as the period from … until …

14 If the sun rises at 0600 UTC and sets at 1800 UTC on a day, a PPL holder without a Night Rating or an Instrument Rating may not take off until …… and must land no later than …… on that day.

15 The additional licence required by a PPL holder to operate a radio-equipped aircraft is known as the …… …………………… Licence, abbreviated ….. . This licence (is/is not) also required for a student pilot to train in a radio-equipped aircraft.

16 A private pilot (without an IMC or Instrument Rating) requires a minimum flight visibility of …. km when flying on a Special VFR Clearance in a control zone.

17 Does a private pilot require a separate rating to fly a microlight aircraft?

18 Is a pilot required by the ANO to record details of each flight in his logbook?

19 Flight time logged should be the time (airborne/from commencing taxiing until stopping after landing/engine start to engine stop).

20 Is a pilot required by the ANO to record the details of every flight simulator session in his logbook? If not, are there any particular flight simulator sessions that must be recorded in his logbook?

21 If a pilot is medically unfit to act as flight crew, the CAA must be advised in writing (as soon as possible/within 7 days/ within 21 days/within 90 days) in the case of injury or pregnancy, and in the case of illness (as soon as possible/within 7 days/within 21 days/within 90 days/as soon as 21 days have elapsed).

22 If a licence holder has a significant injury which makes him unfit to fly, his medical certificate is considered to be suspended immediately. (True/False)?

23 A student pilot or PPL(A) holder may exercise his privileges with a Class 1 or 2 medical certificate. (True/False)?

24 The requirements, standards and privileges laid down for the issue of a medical certificate are contained in the document known as JAR-FCL 3. (True/False)?

25 Following a local anaesthetic (after a visit to the dentist for instance) a pilot (may/ may not) exercise the privileges of his licence for a period of 12 hours after the event.

26 A licence holder (must/must not) seek the advice of an aviation medical examiner (AME) or the CAA medical department as soon as possible if he:

(i) undergoes a surgical operation or invasive procedures;

(ii) is admitted to a hospital or clinic for more than 12 hours;

(iii) requires corrective spectacles or contact lenses on a regular basis;

(iv) needs medication on a regular basis.

27 The privileges of an IMC Rating (may/may not) be exercised outside UK territorial waters without the prior agreement of the State whose airspace it is intended to use.

Exercises 11

Operation of Aircraft

1 May a loaded firearm be carried on an aeroplane?

2 Should a drunk person be carried in an aeroplane?

3 Following the consumption of even just a small amount of alcohol the CAA advises that a pilot should not fly for at least .... hrs. If excessive amounts of alcohol have been consumed a person's ability to fly (may/will not) be significantly impaired for much longer periods.

4 The ANO Article that prohibits drunk persons from aircraft applies to (aircrew only/passengers only/all persons).

5 Is an appropriate radiotelephony operator's licence required by a qualified private pilot to operate the radio in a radio-equipped aircraft?

6 What is the meaning of a series of red flashes directed at an aircraft in flight?

7 Can flying training in an aeroplane occur at any landing area?

8 The commander of an aircraft is required by the ANO to produce his licence and personal flying logbook if so required by an authorised person for up to (1 month/1 year/2 years/3 years/5 years) from the date of the last entry.

9 Unless specified by NOTAM a captive balloon or kite will not be moored within a distance of .... from an aerodrome and be not more than a height above ground level of .... .

10 Is the requirement to carry full documentation waived for an aerial work flight that will remain in UK airspace and the intent is to take off and land at the same aerodrome?

11 Is the requirement to carry full documentation waived for an aerial work flight that will remain in UK airspace and the intent is to take off and land at different aerodromes?

12 Is the requirement to carry full documentation waived for an aerial work flight that will take off and land at the same aerodrome in the UK, but will enter foreign airspace during the flight?

13 Planned parachuting is only permitted with the written permission of the (aircraft owner/aircraft operator/CAA), and then only in such a manner as to not damage people or property, and in an aircraft whose (Certificate of Airworthiness/Maintenance Release to Service/Certificate of Registration) contains a provision that it may be so used.

14 If an aeroplane is expected at an aerodrome the commander must inform the authorities at the aerodrome as soon as possible if the destination is changed or if arrival will be delayed by .... minutes or more. This is to avoid any unnecessary 'overdue action' being taken.

15 When a pilot considers that the safety of his aircraft has been endangered by the proximity of another aircraft, this is known as an ...., and he should:

(a) wait until he lands before reporting the incident to the authorities.

(b) report the incident immediately by radio to the ATSU with which he is in contact, and within 7 days confirm the incident on the appropriate CAA form.

16 The pilot-in-command (shall/shall not) be responsible for the safety of all persons on board during the flight.

17 In the event of an accident involving the aeroplane which results in serious injury or death, the pilot-in-command (shall/may) be responsible for notifying the appropriate authority.

18 The pilot-in-command (must not/should not) begin a flight unless he has ascertained that the aerodrome facilities are adequate for the safe operation of the aeroplane.

19 The pilot-in-command (must not/should not) fly below the operating minima specified for an aerodrome, except with State approval.

20 The pilot-in-command (must/should) ensure that crew members and passengers are briefed on the location and use of oxygen equipment, if fitted.

21 The pilot-in-command (must not/should not) begin a flight unless he is satisfied that the aeroplane is airworthy, registered and has all the appropriate certificates on board.

22 Flights (shall/shall not) be continued towards the planned destination aerodrome unless current weather reports indicate conditions at that aerodrome, or at least one alternate destination aerodrome, are at or above specified minima where the aircraft can divert in the event of deteriorating weather or other contingencies.

23 A flight (may/may not) be conducted in known or expected icing conditions unless the aeroplane is equipped to cope with such conditions.

24 An aeroplane (must/should) be taxied on the movement area of an aerodrome only if the person at the controls has been authorized by the owner, lessee or agent to do so.

25 ICAO recommends that aircraft (should/should not) be refuelled while passengers are leaving the aircraft unless the pilot-in-command or other .... person is present.

26 Placards, listings and instrument markings containing operating limitations prescribed by the State of registry (shall/should) be displayed in the aeroplane.

27 Aeroplanes (must/should) be equipped with a seat or berth for each person on board over a minimum age determined by the State of registry.

28 Aeroplanes operating on VFR flights (must/should) be equipped with an accurate timepiece that indicates the time in hours, minutes and seconds.

29 Mandatory equipment to satisfy the requirements of the Certificate of Airworthiness includes an accessible first-aid kit, a suitable portable fire extinguisher and electrical .... for replacement of those accessible in flight.

30 All single-engined landplanes when flying over water beyond gliding distance from land (must/should) carry one life-jacket or equivalent flotation device for each person on board.

31 Single-engined aeroplanes, when flying over water and more than .... nautical miles from land suitable for an emergency landing, (must/should) be equipped with life-saving rafts capable of carrying all persons on board.

32 Aeroplanes flying over land areas designated by the State as being areas in which search and rescue would be especially difficult (must/should) be equipped with appropriate signalling devices.

33 All aeroplanes on extended flights over water shall be equipped with one life-jacket or equivalent flotation device for each person on board when more than .... nautical miles from land suitable for an emergency landing.

34 In the case of single-engined aeroplanes, what else must be carried when over water and more than 100 nautical miles from land suitable for an emergency landing?

Exercises 12

Distress, Urgency, Safety and Warning Signals (Rule 49)

1 Distress, Urgency or 'lost' calls can be made on the frequency in use or on the emergency service frequency ..... megahertz.

2 Repeated switching on and off of the aircraft landing lights is:
(a) an urgency signal.
(b) a distress signal.
(c) a warning signal.
(d) of no significance.

3 A succession of pyrotechnic single reds fired at short intervals is:
(a) an urgency signal.
(b) a distress signal.
(c) a warning signal.
(d) of no significance.

4 A red parachute flare is:
(a) an urgency signal.
(b) a distress signal.
(c) a warning signal.
(d) of no significance.

5 A pilot should, if he finds himself about to enter, or already in, an active Danger Area, Restricted Area or Prohibited Area, take immediate action:
(a) to avoid the area, or leave it by the shortest route without changing level.
(b) to land as soon as possible.
(c) to climb.

6 If a pilot inadvertently enters a Prohibited Area, but receives no radio instruction, he should ...

7 Mayday repeated 3 times is:
(a) an urgency signal.
(b) a distress signal.
(c) a warning signal.
(d) of no significance.

8 Pan-Pan repeated 3 times is:
(a) an urgency signal.
(b) a distress signal.
(c) a warning signal.
(d) of no significance.

9 The emergency service VHF radio communication frequency 121.5 MHz should be used only for:
(a) urgency calls.
(b) distress calls.
(c) requests for assistance.
(d) all of the above.

10 To indicate an emergency situation to a radar controller, you can squawk Code .... on your transponder, preferably (with/without) Mode C altitude reporting.

11 To indicate a radio failure to a radar controller, you can squawk Code .... on your transponder, preferably (with/without) Mode C altitude reporting.

Exercises 13

Search and Rescue (SAR)

1 SAR stands for ....

2 The emergency service VHF frequency is ..... MHz and is generally reserved for ......, ... and .... calls.

3 A constant SAR Watch over your flight is better achieved when flying in a remote part of Scotland by filing a .... with ATC.

4 Pilots are advised to file a flight plan for SAR purposes when flying over water and more than .... nm from the coast.

5 Pilots (may/may not) file a flight plan for any flight.

6 The pilot of an aircraft not fitted with a radio is advised to file a ....

7 If an aircraft is expected at an aerodrome, the pilot must inform the Air Traffic Service Unit or other Authority at that aerodrome of:
  (a) any change in destination.
  (b) any estimated delay of 45 minutes or more.
  (c) any estimated delay in excess of 1 hour.
  (d) any estimated delay of 3 hours or more.

8 If a pilot lands at an aerodrome other than the intended destination, or decides en route not to land at the originally intended destination, he must ensure that the ATSU at the intended destination is informed within .... minutes of his planned estimated time of arrival there in order to avoid unnecessary action by the alerting services.

9 If an aeroplane fails to arrive at the planned destination, search and rescue action will commence .... minutes after the estimated time of its arrival.

10 To indicate an emergency situation, such as commencing a forced landing due to an engine failure, you should squawk code .... on your SSR transponder.

11 To indicate a need for medical help following a forced landing in a remote area, what symbol should you construct on the ground in a position such that it will be easily visible?

12 To indicate that you have left your aircraft following a forced landing in a remote area and are travelling in a particular direction, what easily visible symbol should you construct on the ground?

13 What squawk code should you set on your SSR transponder to indicate a radio failure?

14 States shall provide search and rescue service within their territories .... hours a day.

15 When an aircraft is believed to be in distress, the rescue coordination centre (shall/should) notify the operator and alert the appropriate ...... and ...... units.

16 If a pilot-in-command observes another aircraft or surface craft in distress, he shall, unless unable or considers it unnecessary, keep in sight the craft in distress for ....

17 A pilot intercepting a distress message may, at his .... while awaiting instructions .... to the position given in the transmission.

18 If the first aircraft to reach an accident scene is not a SAR aircraft it shall take charge of on-scene activities of all other subsequently arriving aircraft until ....

19 What air-to-ground signal should be given during daylight hours to signify that the ground signals have been understood?

20 What air-to-ground signal should be given during darkness to signify that the ground signals have been understood?

Exercises 14

Accident Investigation Regulations

1 A notifiable accident in or over the UK should be reported by the pilot (or operator) to:
  (i) the .... of Air Accidents by the quickest possible means; and
  (ii) the local ....

2 If a person is seriously injured in a taxiing collision, should an 'accident' be notified?

3 If a person is seriously injured while the aeroplane is undergoing maintenance in a hangar, should an 'accident' be notified?

4 If a fuelling truck runs into an unoccupied parked aeroplane, should an accident be notified?

5 If a fuelling truck runs into a parked aeroplane which has just been boarded prior to flight, should an accident be notified?

6 If the engine fails in flight and a safe forced landing on a nearby aerodrome is achieved, should an accident be notified?

7 If the engine fails in flight and a forced landing is achieved in which no person is injured, but one wing is severely damaged, should an accident be notified?

8 If the engine fails in flight and a safe forced landing is achieved in an inaccessible area, should an accident be notified?

9 If the propeller slipstream blows stones back causing a large window to break, but no person is seriously injured, should an accident be notified?

10 If an aircraft part detaches in flight and injures a person on the ground is it considered to be an accident or an incident?

Exercises 15

ICAO Annex Terminology

1 Define the following:
(i) aerodrome beacon
(ii) aerodrome reference point
(iii) aeronautical beacon

2 What does ACAS stand for?

3 Define *aircraft proximity* (AIRPROX).

4 What is a designated area on an aerodrome apron for the parking of aircraft called?

5 Air traffic control service is provided to
(i) expedite the flow of .... and
(ii) prevent .... between aircraft in flight and .... between aircraft and ground ....

6 What is an *airway*?

7 A service which notifies appropriate organisations of aircraft that require search and rescue facilities and other aid (such as for unlawful interference) is known as an ....

8 Define the following:
(i) alert phase
(ii) uncertainty phase
(iii) distress phase

9 Define *altitude*.

10 The *apron* is an area on an aerodrome where aircraft can be ...... for the loading and unloading of ...., ..., or ..... or refuelling or ....

11 What do the letters ATIS stand for?

12 The *ceiling* is the height above ground or water of the lowest layer of cloud below ..... ft covering more than half the sky.

13 Controlled airspace is of defined dimensions within which air traffic control services are provided to ... and ... flights.

14 Take-off run available (TORA) is the length of runway which is declared .... and .... for the ground run of an aeroplane taking off.

15 Take-off distance available (TODA) is the length of the take-off run available plus the ...., if provided.

16 Accelerate-stop distance available (ASDA) is the length of the take-off run available plus the length of the ...., if provided.

17 Landing distance available (LDA) is the length of runway which is declared .... and .... for the ...... run of an aeroplane landing.

18 A flight level is a surface of constant .... pressure which related to a specific pressure datum, 1013.2 mb (hPa), and is separated from other such surfaces by specific ....

19 Ground visibility is visibility at an aerodrome, as reported by a ....

20 Define *identification beacon*.

21 What is IFR flight?

22 A lighter-than-air aircraft is an aircraft that is supported in flight mainly by its .... in the air.

23 Give an example of a lighter-than-air aircraft.

24 The part of an aerodrome used for the taxiing, take-off and landing of aircraft, excluding aprons, is called the ....
..... .

25 What is a NOTAM?

26 Radar is a radio detection system which provides information on range, ... and .... of objects.

27 Define radar vector.

28 A serious injury is an injury sustained by a person in an accident which requires hospitalisation for more than .... hours (from within 7 days of the accident).

29 What is a *signal area*?

30 What normally indicates the beginning of the threshold of a runway?

31 The transition layer is the airspace between the transition .... and transition ....

32 What is VMC?

Aviation Law, Flight Rules and Procedures

Answers 1

Aviation Law and Legislation

1 1944, Montreal
2 Articles
3 Annexes
4 has
5 is
6 will
7 must
8 do
9 does
10 may
11 do
12 may not
13 shall
14 is
15 will
16 shall
17 shall
18 must
19 shall
20 must
21 shall
22 shall
23 may not
24 may
25 shall
26 may
27 International Civil Aviation Organization
28 ENR section of the UK Aeronautical Information Publication
29 Notice to Airmen
30 Aeronautical Information Circular, pink
31 8 hours
32 3 minutes
33 2 minutes
34 8 nm, 4 minutes
35 well before
36 no
37 no
38 yes

39 20%, 10%
40 20%
41 25% or more
42 red
43 black
44 turbine fuel
45 AIC 12/1996 (Pink 120)
46 Air Navigation Order
47 Parts, Articles

Answers 2

Rules of the Air

1 (a)
2 all
3 the pilots
4 right
5 right
6 should
7 over, under or crossing ahead
8 exists
9 must, must
10 has, does not have
11 has
12 right
13 left
14 70°
15 (c)
16 (a)
17 should
18 left
19 8 nm, 4 nm
20 has
21 lower
22 should not
23 right
24 left
25 left
26 right
27 left
28 red
29 green

30 110°

31 white, 70°

32 30, 30

33 does not

34 does

35 does not

36 (a)

37 (a)

38 (b)

39 are required to

40 may

41 airship, the airship

42 500 ft

43 should

44 1,500 ft, 600 metres, government, CAA, licensed

45 1,000 metres, 1,000 people

46 over a congested area such as a city, town or settlement

47 must

48 1,500 ft, 600 metres

49 1,000 metres

50 (a)

51 must, safety pilot

52 may

53 are

54 must

55 may, 10 seconds, red and green

56 30, 30, end, beginning

Answers 3

Aerodromes

1 yes

2 yes

3 yes

4 emergency, prior permission

5 government, CAA, aerodrome licensed for training

6 yes

7 mandatory

8 yellow

9 yes

10 the holding point for a runway which must not be crossed without permission from ATC

11 no

12 two, red

13 two, green

14 green and white, or white and white

15 right

16 turn left and keep the other aircraft on your right

17 right

18 true

19 the aeroplane under tow

20 aircraft taking off or landing, vehicles towing aircraft, aircraft taxiing, vehicles

21 2,000 ft aal, 2 nm

22 2,000 ft aal, 2.5 nm

23 (a)

24 (c)

25 (b)

26 take off or land on a runway, but taxiing is not confined to hard surfaces

27 (a)

28 (d)

29 the direction of take-off and the direction of landing may not be the same

30 the section between the crosses is unfit for the movement of aircraft

31 aircraft may move on the manoeuvring area and apron only with ATC permission

32 (b)

33 (b)

34 no

35 should

36 yes

37 gliding is taking place at the aerodrome

38 white

39 true

40 should, yellow cross

41 red flag for left-hand, green flag for right-hand

42 stop

43 clear to move on the manoeuvring area and apron

44 return to the aerodrome and wait for permission to land

45 you may land

46 do not land; aerodrome not available for landing

47 give way to other aircraft and continue circling

48 do not land; wait for permission

49 return to starting point on aerodrome

50 land at this aerodrome after receiving a continuous green light, and then, after landing and after receiving green flashes, proceed to the apron

51 (a)

52 (c)

53 park in this bay

54 (d)

55 stop urgently

56 (c)

57 (a)

58 (c)

59 cross your hands in front of your face, palms facing outwards, and then move arms outwards

60 extend your arms with the palms facing outwards, then move your hands inwards to cross in front of your face

61 (a), (b) & (c)

62 red

Answers 4

Altimeter-Setting Procedures

1 decreases

2 changes

3 is

4 QNH

5 altitude

6 height

7 elevation

8 QFE

9 1013

10 FL35

11 FL245

12 transition altitude

13 transition level

14 terrain and other traffic

15 aerodrome QNH or QFE

16 regional QNH, also known as the regional pressure setting

17 (b)

18 (b)

19 (d)

20 refer to the text

21 (c)

22 (d)

23 magnetic track

24 (a)

25 (b)

26 (b)

27 (c)

28 4,010 ft

29 5,600 ft

30 aerodrome QNH

31 aerodrome QFE, aerodrome QNH

Answers 5

Airspace

1 London and Scottish, FIRs

2 24,500 feet, FL245

3 upper airspace, UIR, Upper Information Region

4 uncontrolled or controlled

5 A to E, F and G

6 B, controlled

7 is not

8 A, controlled, is not

9 D, E

10 F or G, F, G

11 will

12 Advisory Route, ADR, 10 nm wide

13 G, Open-FIR

14 Aerodrome Traffic Zone

15 see our text

16 position and height

17 Military Aerodrome Traffic Zone

18 see our text

19 does

20 aerodrome QFE

21 (a)

22 5 nm

23 CTR, ground level

24 CTA, altitude or flight level

25 Airway

26 Terminal Control Area, TMA

27 one VHF radio

28 5 nm, has

29 may

30 may not

31 (b)

32 may
33 may
34 Prohibited Area
35 Restricted Area
36 Danger Areas, solid red
37 Danger Areas, pecked red
38 Danger Area, 57°N–58°N, 15,000 ft amsl
39 Danger Area, 55°N–56°N, 1,000 to 4,500 ft amsl
40 300 ft agl
41 500 ft agl
42 300 ft agl, 500 ft agl
43 heights up to the cloud base, or the base of controlled airspace, whichever is the lower
44 heights up to the cloud base or the base of controlled airspace, whichever is the lower
45 2,000 ft agl
46 2,000 ft agl
47 1,500 ft agl
48 2,000 ft agl, between 250 ft and 500 ft agl
49 are, Areas of Intense Aerial Activity
50 second
51 should, should
52 2,000 ft of cable, altitude 10,000 ft amsl, are
53 Temporary Class A airspace, are
54 FL150

Answers 6

Air Traffic Services

1 collisions, collisions, manoeuvring, maintain, flights, organisations, search
2 (a) IFR
(b) IFR and VFR
(c) ATC
(d) IFR, VFR
(e) ATC, far as is practical
(f) air traffic, if requested
(g) VFR, IFR, flight, if requested
3 Air Traffic Control
4 Aerodrome Control and Approach Control
5 lamp or pyrotechnic signals, ground signals
6 Flight Information Service
7 does not
8 does not
9 ATC
10 ATC

11 A/G (Air/Ground)
12 FIS
13 radio
14 should not
15 is
16 both pilots
17 all of them
18 'book out'
19 10 nm
20 need not
21 must
22 must
23 need not but should
24 30 minutes
25 30 minutes
26 self-briefing, does not
27 4
28 2 hours
29 is
30 (a)
31 Facilitation

Answers 7

Visual Flight Rules (Rules 24–27)

1 Visual Meteorological Conditions
2 is not
3 D, E
4 F, G
5 8 km, 1,500 metres, 1,000 ft
6 5 km
7 uncontrolled
8 F and G, 5 km, clear of cloud and in sight of the surface
9 1,500 metres, may
10 both
11 Rules of the Air
12 ANO Schedule 8
13 3 km, clear of cloud, in sight of the surface
14 no
15 may not, A
16 Special VFR
17 a Special VFR clearance
18 is
19 is not
20 absolves
21 is not

22 the pilot
23 (a)
24 true
25 may, must
26 the time between 30 minutes after sunset and 30 minutes before sunrise, as measured at surface level
27 true
28 1830 UTC
29 are
30 FL45
31 FL55 (magnetic track 090°M)

Answers 8
Instrument Flight Rules (Rules 28–32)

1 Instrument Flight Rules
2 1,000 ft within 5 nm
3 3,000 ft amsl, clear of cloud and in sight of the surface
4 (d)
5 magnetic track
6 FL75, 1013
7 FL65 (based on magnetic track 273°M), no change since amount of drift only determines the heading required to achieve the desired track
8 FL80

Answers 9
Registration and Airworthiness

1 a group of letters and/or numbers
2 lower, left
3 50
4 fireproof metal or other suitable fireproof material – secured near the main entrance
5 aeroplanes yes, gliders no
6 true
7 is
8 3 years
9 short take-off and landing (STOL)
10 true
11 Technical Log
12 (c)
13 will, Release
14 2,730 kg, Private or Special
15 yes

16 yes
17 Aircraft Weight Schedule
18 insurance policies, warranties and guarantees
19 2 years
20 (b)
21 6 months
22 the State of registry
23 all limiting mass (weights), centres of gravity position, mass (weight) distributions and floor loadings
24 structural, flying
25 crew and payload, usable fuel and drainable oil

Answers 10
Pilots' Licences

1 17 years
2 true
3 (b)
4 24 months
5 12 hours, 6 hours PIC, 12 take-offs and landings, 3 months
6 a valid medical certificate, and either a valid CofT or CofR for the type of aircraft to be flown
7 24 months
8 false
9 Air Navigation Order (ANO)
10 3 km
11 3 km
12 three, one, one
13 from 30 minutes after sunset until 30 minutes before sunrise
14 0530 UTC, 1830 UTC
15 Flight Radiotelephony Operator's Licence, FRTO, is not
16 10 km
17 no, he requires a separate licence (PPL(A) Microlight)
18 yes
19 from commencing taxiing until stopping after landing
20 no; yes – particulars of flight simulator tests
21 injury and pregnancy: as soon as possible; illness: as soon as 21 days have elapsed
22 true

23 true
24 true
25 may not
26 must
27 may not

Answers 11

Operation of Aircraft

1 no
2 no
3 8 hours, may be impaired
4 all persons
5 yes
6 do not land, aerodrome not available for landing
7 yes, except when the training is for the purpose of qualifying for a pilot's licence, aircraft rating or night rating
8 2 years
9 5 km, 60 metres
10 yes
11 no
12 no
13 CAA, Certificate of Airworthiness
14 45 minutes
15 Airprox, (b)
16 shall
17 shall
18 must not
19 must not
20 must
21 must not
22 shall not
23 may not
24 must
25 should not, qualified
26 shall
27 must
28 must
29 spare fuses
30 must
31 100, must
32 must
33 50
34 Rafts capable of carrying all persons on board, appropriate life-saving equipment

and equipment for making pyrotechnic distress signals

Answers 12

Distress, Urgency, Safety and Warning Signals (Rule 49)

1 121.5 MHz
2 (a)
3 (b)
4 (b)
5 (a)
6 leave the Prohibited Area as quickly as possible without descending
7 (b)
8 (a)
9 (d)
10 7700, with
11 7600, with

Answers 13

Search and Rescue (SAR)

1 Search and Rescue
2 121.5 MHz; "Mayday" repeated three times (distress); "Pan-Pan" repeated three times (urgency) or requests for assistance (e.g. 'lost')
3 flight plan
4 10 nm
5 may
6 flight plan
7 (a) and (b)
8 30 minutes
9 30 minutes
10 7700
11 a large ✗
12 a large →
13 7600
14 24
15 shall, search, rescue
16 as long as necessary
17 discretion, proceed
18 the first SAR aircraft arrives
19 rocking the aircraft's wings
20 flashing on and off twice the aircraft's landing lights, or if not so equipped, switching on and off twice its navigation lights

Answers 14

Accident Investigation Regulations

1 Chief Inspector of Air Accidents, local Police Authority
2 yes
3 no
4 no
5 yes
6 no
7 yes
8 yes
9 no
10 accident

Answers 15

ICAO Annex Terminology

1 (i) an aeronautical beacon used to indicate the location of an aerodrome from the air
(ii) the designated geographical location of the aerodrome
(iii) an aeronautical ground light visible from all directions, either continuously or intermittently, to indicate the location of a particular point on the surface of the earth
2 airborne collision avoidance system
3 a situation where minimum safe separation distances between aircraft in flight have been exceeded
4 aircraft stand
5 traffic, collisions, on the ground, obstructions
6 a corridor-shaped control area equipped with radio navigation aids
7 alerting service
8 (i) where concern is registered regarding the safety of an aircraft and its occupants
(ii) when the safety of an aircraft and its occupants is uncertain
(iii) where it is believed that an aircraft and its occupants require immediate assistance or are threatened by grave or imminent danger.
9 the vertical distance of a point from mean sea level
10 parked, passengers, mail, cargo, maintenance
11 automatic terminal information service
12 20,000
13 IFR, VFR
14 available, suitable
15 length of the clearway
16 stopway
17 available, suitable, ground
18 atmospheric, pressure intervals
19 meteorological observer
20 an aeronautical beacon that flashes a coded signal such that its location can be identified
21 a flight made under Instrument Flight Rules
22 buoyancy
23 a hot-air balloon
24 manoeuvring area
25 Notice to Airmen, a document which contains urgent information concerning the establishment, condition or change in any aeronautical facility, service, procedure or hazard
26 position, elevation
27 a heading instruction given to a pilot by a radar controller, based on radar information
28 48
29 an area on an aerodrome used for the display of ground signals
30 'piano key' markings
31 altitude, level
32 Visual Meteorological Conditions

Meteorology

Exercises 16

The Atmosphere

1   The atmosphere extends further into space above the (equator/poles).

2   The region of the atmosphere closest to the earth and in which 'weather' occurs is called the ....

3   The second layer of the atmosphere is called the .... and the boundary between it and the troposphere is called the ....

4   The main gases that form the atmosphere are ...

5   Most of the water vapour in the atmosphere is contained in the:
(a)   tropopause.
(b)   troposphere.
(c)   stratosphere.

6   Air density generally (increases/decreases/stays the same) as altitude is gained.

7   At sea level, air density is approximately:
(a)   1,225 grams per cubic metre.
(b)   1013 mb (hPa).
(c)   29.6 inches of mercury.

8   Temperature generally (increases/decreases/stays the same) as altitude is gained.

9   There is marked vertical movement of air in the troposphere. (True/False?).

10   A body of air over an ocean is referred to as:
(a)   maritime air.
(b)   continental air.
(c)   polar air.
(d)   oceanic air.

11   An air mass that passes over an ocean is likely to be (more/less) moist than an air mass that passes over a continent.

Exercises 17

Heating Effects in the Atmosphere

1   The air surrounding the earth is mainly:
(a)   heated directly by the sun.
(b)   heated from below by the earth's surface.

2   Heating is greatest in the:
(a)   tropics.
(b)   temperate zones.
(c)   polar regions.

3   Hot air (rises/sinks).

4   Cool air (rises/sinks).

5   Terrestrial radiation is:
(a)   the direct heating of the earth by the sun.
(b)   the re-radiation of heat from the earth.

6   Solar heating of the earth occurs:
(a)   only by day.
(b)   continually.

7   The sea heats (more/less) rapidly than land.

8   The sea cools (more/less) rapidly than land.

9   Generally the sea is (warmer/cooler) by day than the land.

10   Generally the sea is (warmer/cooler) by night than the land.

11   Water has a (higher/lower) specific heat than land.

12   Cloud coverage (reduces/increases/does not affect) the heating of the earth's surface.

13 Cloud coverage (reduces/increases/does not affect) the cooling of the earth's surface by the terrestrial re-radiation of heat.

14 The transfer of heat as electromagnetic waves is called the process of ....

15 The transfer of heat from body to body is called the process of ....

16 The transfer of heat by the horizontal motion of an air mass is called ....

17 The transfer of heat by the vertical motion of an air mass is called ....

18 A sea breeze blows (offshore/onshore) during the late afternoon.

19 The land breeze blows (offshore/onshore) at dawn.

20 The wind that flows down mountain slopes at night due to cooling is called a .... wind.

21 The wind that flows up mountain slopes by day caused by heating is called an .... wind.

22 If the air at the earth's surface is cooler than that above, a .... is said to exist.

23 Convert 10°C to °F.

24 Convert −15°C to °F.

25 Convert 41°F to °C.

26 An inversion means that the temperature (increases/decreases/stays constant) as height increases.

Exercises 18

Atmospheric Pressure

1 Atmospheric pressure (increases/decreases/stays the same) as height is gained.

2 Pressure drops by about .... mb per 30 ft gain in height in the lower levels of the atmosphere.

3 The daily heating and cooling effects cause the .... variation of pressure.

4 A line on a map joining places of equal sea level pressure is called:

(a) an isobar.

(b) an isopress.

(c) a pressure gradient.

(d) an equi-pressure line.

5 The variation of pressure with horizontal distance is called the ....

6 There is a natural tendency for air to flow from areas of .... pressure to areas of ... pressure.

7 An aeroplane, flying so that the altimeter indicates 2,500 ft with the current regional QNH set on the subscale, is flying towards an area of *low* pressure. If the pilot fails to revise the subscale setting as the QNH changes, then the aeroplane will:

(a) gradually descend.

(b) gradually climb.

(c) maintain 2,500 ft amsl.

8 Unless the pilot periodically resets the lower regional QNH as an aeroplane flies towards an area of low pressure, the altimeter will (over/under)-read, and the aeroplane will be (lower/higher) than the altimeter indicates.

Exercises 19

The International Standard Atmosphere (ISA)

1 ISA stands for the ....

2 Temperature at sea level in the ISA is ....

3 Pressure at sea level in the ISA is ....

4 Density at sea level in the ISA is ..... .

5 Temperature in the ISA decreases by ....°C for each 1,000 ft gained in the lower levels of the atmosphere.

6 The actual rate of decrease of temperature with altitude is called the ....
....

7 Above approximately 36,000 ft in the theoretical International Standard Atmosphere, the temperature ceases to fall and remains constant at approximately (0°C/+57°C/–57°C/+57°F/ –57°F).

8 The temperature at 3,000 ft amsl is given as +6°C. The difference compared with the International Standard Atmosphere (ISA) is ...?

Exercises 20
Wind

1 The horizontal flow of air is called ..... .

2 Meteorologists relate wind direction to (true/magnetic) north, and so all winds that appear on meteorological forecasts or observations are in (°T/°M).

3 Runways are described in terms of their (true/magnetic) direction, and so any winds passed to the pilot by the Tower for the purpose of taking off or landing are in (°T/°M).

4 280/34KT on a meteorological forecast or observation means a wind of strength of .... blowing from a direction of ....°T.

5 A wind of 270/25 is passed to the pilot of an aeroplane on approach to land by the Tower. The wind direction is expressed (°M/°T) and its strength is 25 (knots/mph/kph/metres per second).

6 A wind whose direction has changed in a clockwise direction has ....

7 A wind whose direction has changed in an anticlockwise direction has ....

8 A wind changing from 280/12 to 340/ 18 has (veered/backed).

9 The force that causes a parcel of air to start moving from an area of high pressure to an area of low pressure is called the .... force.

10 The apparent force that causes the curving direction change of the wind is called the .... force.

11 The Coriolis force is caused by the .... of the earth and is greatest near the (poles/equator/temperate zones).

12 The faster the airflow, the (greater/less) the Coriolis effect.

13 In the northern hemisphere, the Coriolis force curves the airflow to the (right/ left).

14 If the pressure gradient force is balanced by the Coriolis force so that the wind blows parallel to the isobars, then this wind is called the .... wind.

15 The geostrophic wind has the low-pressure area on its (left/right) in the northern hemisphere.

16 State Buys Ballot's law for the northern hemisphere.

17 If an aircraft (in the northern hemisphere) is experiencing right drift (i.e. the wind is from the left), it is flying towards a region of (high/low) pressure.

18 If the aeroplane is experiencing left drift (i.e. the wind is from the right), it is flying towards a region of (high/low) pressure.

19 In the northern hemisphere, when flying at a constant indicated altitude towards an area of low pressure, an aeroplane will experience (left/right/no) drift and, unless the subscale is periodically reset, the altimeter will read (too high/too low/correctly).

20 For the wind to blow anticlockwise around a low-pressure system, the pressure gradient force will (exceed/be less than) the Coriolis force.

21 For the wind to blow clockwise around a high-pressure system, the pressure gradient force will (exceed/be less than) the Coriolis force.

22 The wind that flows around curved isobars is called the:

(a) curved wind.

(b) geostrophic wind.

(c) gradient wind.

(d) isobaric wind.

23 Surface wind is measured at .... metres above ground level.

24 Compared to the wind at altitude, the surface wind is (increased/decreased) in strength by the effects of ....

25 Compared to the gradient wind at altitude, the surface wind will (back/veer).

26 The backing of the surface wind is more pronounced over (land/sea).

27 The surface wind resembles the gradient wind at altitude more closely by (day/night).

28 There is (more/less) vertical motion in the atmosphere by day than by night.

29 The surface wind by day is generally (stronger/weaker) than the surface wind by night.

30 Flight at low level in strong winds is likely to be (more/less) turbulent over the land than over the sea.

31 If the wind at altitude is 240/35, the most likely wind on the ground is:

(a) 270/20.

(b) 270/40.

(c) 220/40.

(d) 220/20.

32 Strong air flows associated with a cumulonimbus cloud will occur (within/in the vicinity of) the cloud.

33 If the surface wind is 330/20, the wind at 2,000 ft is likely to be:

(a) 350/30.

(b) 310/30.

(c) 350/15.

34 The variation of wind speed and/or direction from place to place is called ....

35 Low-level windshear (is possible/will never occur) when an aeroplane flies through an inversion layer.

36 A strong wind flow over a mountain range will cause strong downcurrents and turbulence on the (windward/lee) side of the mountains.

37 Lenticular (lens-shaped) clouds may form well above the mountains as a result of strong winds causing m....... w.... or s........ w..... .

38 Strong westerly winds across a north-south mountain range may cause strong and possibly hazardous downdrafts to the (north/south/east/west) of the mountain range.

39 The ability of an aeroplane to climb is degraded when flying up valleys towards high ground in a strong (headwind/tailwind).

40 A good indication that mountain waves are present is the formation of (stratus/cumulus/lenticular) clouds.

41 When a strong wind flows over a mountain range, there may be strong downdrafts on the (windward/lee) side, which an aeroplane may not be able to outclimb.

42 The effect of mountain waves may sometimes be felt as far as (1/5/20/40) nm downwind from the mountains that cause them.

Exercises 21

Cloud

1 Name the four main families of clouds.

2 'Lumpy' or 'heaped' clouds belong to the .... family.

3 An extensive layer of cloud belongs to the .... family.

4 Dense, white clouds resembling a 'cauliflower', and from which showers are falling, are called .... clouds.

5 A heavy, dense cloud with associated thunder and lightning is a .... cloud.

6 A rain-bearing cloud may have the word (stratus/nimbus/cumulus/cirrus) associated with its name.

7 As water vapour condenses to form liquid water, it (absorbs/gives off) latent heat.

8 The amount of water vapour carried in a parcel of air is called ....

9 Warm air can hold (more/less/the same amount of) water vapour compared to cold air.

10 The greater the amount of water vapour in a parcel of air, the (higher/lower) its dewpoint temperature.

11 As a parcel of air cools, its ability to hold water vapour (increases/decreases/remains unaltered).

12 If a parcel of air cools to the particular temperature where it is carrying the maximum amount of water vapour that it can, then it is said to be ....

13 The percentage of water vapour in the air compared to what it is capable of carrying at that temperature is called its ....

14 When a parcel of air is saturated, its relative humidity is:
(a) 100%.
(b) 0.

15 As a parcel of air cools, its relative humidity (increases/decreases/stays the same).

16 The temperature at which a cooling parcel of air reaches saturation is called its:
(a) saturation temperature.
(b) dewpoint temperature.
(c) moisture temperature.
(d) cooling temperature.

17 As air temperature cools to the dewpoint temperature, the relative humidity:
(a) falls.
(b) remains constant.
(c) rises.
(d) rises to 100%.

18 The closer the actual air temperature is to the dewpoint temperature, the (higher/lower) the relative humidity, and the (closer to/further from) saturation is the parcel of air.

19 If the air continues to be cooled below the temperature at which it reaches saturation (i.e. below its dewpoint temperature), then the water vapour will ...

20 As air expands, it (warms/cools).

21 A process in which heat is neither added nor subtracted is called an .... process.

22 As a parcel of air rises and cools, its relative humidity (increases/decreases/stays the same).

23 Is cumulus cloud formed by an adiabatic process as air rises?

24 Unsaturated (or 'dry') air cools adiabatically as it rises by about ....°C/1,000 ft gain in height. This is known as the ... . lapse rate.

25 After the air is cooled to its dewpoint, the water vapour starts to condense into liquid water and so a ..... is formed.

26 As water vapour condenses into liquid water it (absorbs/gives off) latent heat, which (increases/decreases) the rate at which saturated air cools as it rises.

27 The saturated adiabatic lapse rate is approximately (one-half/one-third/the same as) the dry adiabatic lapse rate. The SALR is approximately ...°C/1,000 ft.

28 The rate of change of temperature in the surrounding air that is not rising is called the ....

29 The level at which the cloud base forms depends upon:
  (a) the moisture content of the cloud and its dewpoint temperature.
  (b) the temperature of the environment.

30 If the moisture content of a parcel of air is such that its dewpoint temperature is +7°C, at what height above the ground is its base likely to form if the surface air temperature is +16°C?

31 If the moisture content of a parcel of air is such that its dewpoint temperature is +7°C, at what height above the ground is its base likely to form if the surface air temperature is +19°C?

32 If the moisture content of a parcel of air is such that its dewpoint temperature is +7°C, at what height above the ground is its base likely to form if the surface air temperature is +22°C?

33 If the moisture content of a parcel of air is such that its dewpoint temperature is +13°C, at what height above the ground is its base likely to form if the surface air temperature is +22°C?

34 Cloud formed by turbulence and mixing is known as .... cloud.

35 Cloud formed by a mountain range causing the uplift of air is called .... cloud.

36 Orographic uplift of unstable air is more likely to cause the formation of (stratiform/cumuliform) cloud.

37 Orographic uplift of stable air is more likely to cause the formation of (stratiform/cumuliform) cloud.

38 If moist air flows up and over a mountain range, forming cloud and causing rain, the wind on the lee side of the mountain range will be (warmer/cooler/the same temperature) and (drier/wetter/the same humidity). This is called the ..... wind effect.

39 As air flows up a mountain range its temperature will fall at about:
  (a) ....°C/1,000 ft when it is unsaturated; and
  (b) ...°C/1,000 ft once it has cooled sufficiently to become saturated.

40 Precipitation consisting of water drops is called ..... .

41 Precipitation consisting of small balls of ice is called ..... .

42 Precipitation consisting of branched and star-shaped ice crystals is called .....

43 Showers generally fall from (cumuliform/stratiform) clouds.

44 Drizzle generally falls from (cumuliform/stratiform) clouds.

45 Rain which falls from the base of clouds but which evaporates before reaching the ground is called ....

46 The cloud associated with standing waves (or mountain waves) is known as .... cloud.

Exercises 22

Thunderstorms

1 List three conditions necessary for a thunderstorm to develop.

2 List the three stages of a thunderstorm's life cycle.

3 Specify three types of 'trigger actions' that can lead to the formation of a thunderstorm in unstable, moist air.

4 At what stage in the life of a typical thunderstorm are there strong warm updrafts over a diameter of 1 or 2 nm with no significant downdrafts?

5 The temperature in a forming cumulonimbus cloud is (higher than/lower than/the same as) the outside environment.

6 The beginning of what stage in the life of a typical thunderstorm is signalled by the first lightning flashes and the first rain from the cloud base?

7 In the mature stage of a typical thunderstorm there (will be/will not be) both strong updrafts and downdrafts inside the cloud, and there will be very strong (warm/cool) downdrafts flowing from the base of the cloud.

8 If the top of a cumulonimbus cloud spreads out, it is referred to as an ....., and is a sign that the cloud is well developed.

9 There are only updrafts inside a cumulonimbus cloud. (True/False)?

10 A thundercloud is a hazard to aviation:
   (a) only within the cloud.
   (b) only within and directly under the cloud.
   (c) within about 10 nm.

11 List four of the hazards to aviation caused by thunderstorms.

Exercises 23

Air Masses and Frontal Weather

1 An air mass that has passed over an ocean is referred to as .... air.

2 An air mass that has passed over a large land mass is referred to as .... air.

3 As cool air travels across warm land it absorbs heat and becomes (stable/unstable).

4 Warm air moving across a cooler surface will lose heat and become more (stable/unstable).

5 Tropical air moving north over the oceans until it reaches the UK will be (stable/unstable) and (dry/moist).

6 The slow sinking of an upper air mass is called ....

7 The slow rising of a large air mass is associated with (convergence/divergence) at the earth's surface.

8 Convergence at the earth's surface is associated with a (high/low)- pressure system.

9 Divergence at the earth's surface is associated with a (high/low).

10 Subsiding air becomes (warmer/cooler), (moister/drier) and more (stable/unstable).

11 Subsidence is associated with (stability/instability).

12 Convergence is associated with (stability/instability).

13 A low-pressure system is associated with (stability/instability).

14 Warm air displacing cold air at the surface is called a ....

15 As a warm front approaches, the cloud base (lowers/rises).

16 The general cloud associated with a warm front is (stratiform/cumuliform).

17 If an aeroplane takes off at an aerodrome prior to the passage of a warm front, it will be in the (warm/cold) air mass.

18 In a warm front, the warm air at altitude (precedes/follows) the passage of the front at the surface.

19 The slope of a typical warm front is 1 in (10/50/150/1,000).

20 As a warm front approaches, the first sign could be high level (cirrus/stratus/cumulonimbus) cloud about (20/200/600/2,000) nm ahead of the surface front.

21 Rain associated with a warm front may fall:

(a) only after the surface front has passed.

(b) only in a narrow band some 10 nm wide near the surface front.

(c) up to several hundred miles ahead of the surface front.

22 List the five general characteristics of a warm front.

23 Cold air displacing warm air at the surface is called a ... .

24 Clouds associated with a cold front is of the (cumuliform/stratiform) type.

25 The change in weather with the passage of a cold front (may/will not) be quite sudden.

26 Precipitation associated with a cold front is most likely to be (drizzle/showers).

27 As a cold front passes, the wind will (veer/back).

28 As a cold front passes, the air temperature will (rise/fall), and the dewpoint temperature will (rise/fall).

29 Visibility following the passage of a cold front is likely to be:

(a) poor in continuous drizzle.

(b) excellent in all directions.

(c) good, except in showers.

30 The slope of a typical cold front is 1 in (10/50/150/1,000).

31 As a cold front passes, the temperature will (rise/fall).

32 As a warm front passes, the wind will (veer/back).

33 As a warm front passes, the temperature will (rise/fall).

34 As a warm front passes, the air temperature will (rise/fall), and the dewpoint temperature will (rise/fall).

35 A wind of 20 knots from the west does not allow you to take off on the north-south runway at Sleap in your particular aeroplane because the maximum cross-wind limit is exceeded. If a cold front passes, it is likely that you will be able to take off into the (north/south).

36 Rain is more likely to precede the passage of a warm front than a cold front. (True/False)?

37 The general visibility away from the clouds and showers associated with a cold front will be (better/poorer) than the general visibility associated with a warm front.

38 If a cold front overtakes a warm front, the result is an .... front.

39 There (may/will never) be intense weather associated with an occluded front.

40 Flying into an area where the cloud base is lowering to within 1,000 ft of the terrain, ice starts to form on the wings. Your best course of action is to:

(a) climb, even though it means entering cloud.

(b) descend into warmer air, but continue on.

(c) maintain track and level.

(d) turn back.

41 In the northern hemisphere, wind flows ....wise around a low.

42 Flying towards a low an aeroplane will experience ..... drift.

43 In the northern hemisphere, wind flows .....wise around a high.

44 Flying towards a high in the northern hemisphere, an aeroplane will experience .... drift.

45 (Convergence/divergence) at the surface is associated with a low.

46 (Rising/sinking) air is associated with a low.

47 A V-shaped extension of low pressure is called a ....

48 A cold front usually moves (faster/slower) than a warm front.

49 Highs generally have a (weaker/stronger) pressure gradient than lows.

50 (Convergence/divergence) at the surface is associated with a high.

51 (Rising/subsiding) air is associated with high-pressure systems.

52 Subsiding air is very (stable/unstable).

53 Another name for the high is the ....

54 A U-shaped extension of isobars surrounding a high is called a ....

55 An area of almost constant pressure located between two highs and two lows is called a .... .

56 If the wind veers and rain/drizzle that has been falling steadily for many hours gradually ceases, then it is possible that a (warm/cold) front has passed.

57 If an aeroplane is experiencing right (starboard) drift in the northern hemisphere and the pilot fails to periodically revise the regional QNH set in the altimeter subscale, then the aeroplane will gradually (climb/descend).

Exercises 24

Icing

1 Water may freeze when the temperature is less than .... °C.

2 Ice that forms on the wings, fuselage, propeller, etc., is known as .... ice.

3 Ice that forms in the carburettor is known as .... ice, which (can/cannot) form if the outside air temperature is as high as +25°C.

4 The most dangerous form of airframe icing is:
(a) clear ice.
(b) hoar frost.
(c) dry ice.
(d) rime ice.

5 Airframe icing (may/will not) be more severe at an air temperature of −3°C than at −40°C.

6 Sometimes water exists as liquid even though its temperature is less than the freezing point of water. Such droplets are said to be ....

7 Large supercooled water drops striking a cold airframe are likely to form:
(a) clear ice.
(b) hoar frost.
(c) dry ice.
(d) rime ice.

8 Very small supercooled water droplets striking a cold airframe are likely to form:
(a) clear ice.
(b) hoar frost.
(c) dry ice.
(d) rime ice.

9 For carburettor ice to form, the outside air must be:
(a) below freezing.
(b) moist.
(c) dry.
(d) cold and moist.

10 Frost or ice on the leading edge of a wing (is/is not) dangerous.

11 Ice of any type should be cleared off the aeroplane before flight. (True/False)?

12 If ice forms over the static vent of an aeroplane and blocks it during the climb, the altimeter will read:

(a) zero.

(b) a constant altitude.

(c) correctly.

13 If ice forms over the static vent of an aeroplane and blocks it during the climb, the vertical speed indicator will read:

(a) zero.

(b) a constant altitude.

(c) correctly.

14 If ice forms over the static vent of an aeroplane and blocks it during the climb, the airspeed indicator will read:

(a) zero.

(b) too fast.

(c) too slow.

(d) correctly.

15 Carburettor ice is most likely to form if the air temperature and dewpoint temperature are respectively:

(a) −10°C and −20°C.

(b) +12°C and +10°C.

(c) +12°C and 0°C.

(d) 0°C and −6°C.

16 Throttle icing is more likely at (high/ low) power settings.

17 When flying in the cold sector underlying the warmer air in a warm front there is:

(a) no possibility of airframe ice forming.

(b) no possibility of airframe ice forming if the temperature of the rain falling out of the warm sector is above 0°C.

(c) a possibility of clear ice forming in rain.

(d) rarely any rain.

Exercises 25

Visibility

1 When approaching to land, (slant/vertical/horizontal) visibility is most important to a pilot.

2 Small liquid droplets suspended in the air that reduce the visibility to less than 1 km is called .... .

3 Small liquid droplets suspended in the air that reduce the visibility to say 1.5 km is called ..... .

4 Terrestrial re-radiation causes the surface of the earth to cool at night. It, in turn, cools the air in contact with it and .... fog may form.

5 List three requirements for radiation fog to form.

6 If a warm, moist air mass flows over a cold surface, it will cool. If it cools to its dewpoint, then fog will form. This is known as .... fog.

7 A warm maritime airflow over land may give rise to .... fog.

8 Fog formed by the interaction of two air masses is called .... fog.

9 If the air temperature increases with height, then an .... exists.

10 It is more likely that an inversion will exist at (dawn/dusk).

11 If radiation fog forms on a clear night with light winds, an increase in wind strength from 5 kt to 18 kt:

(a) will change the radiation fog to advection fog.

(b) may cause the fog to lift and become low stratus.

(c) will have no effect.

12 An early morning fog over the sea lasts all day. As the land heats up, the sea fog:

(a) may drift in over the land.

(b) will always disperse.

(c) will always remain over the sea.

13 Visibility will be (greater/poorer) when flying into the sun than when flying 'down-sun'.

14 On a cool and cloudless night with no wind, and the air in contact with the surface cooled to its dewpoint temperature (say +5°C), which of the following is most likely to form?
(a) Dew.
(b) Frost.
(c) Mist.
(d) Fog.
(e) Stratus.

15 On a cool and cloudless night with no wind, and the air in contact with the surface cooled to its dewpoint temperature (say −5°C), which of the following is most likely to form?
(a) Dew.
(b) Frost.
(c) Mist.
(d) Fog.
(e) Stratus.

16 On a cool and cloudless night with a light wind, and the very moist air in contact with the surface cooled to its dewpoint temperature (say +7°C), which of the following is most likely to form?
(a) Dew.
(b) Frost.
(c) Mist.
(d) Fog.
(e) Stratus.

17 If warm maritime air flows over a cold land surface, it may form:
(a) radiation fog.
(b) frontal fog.
(c) advection fog.
(d) hail.

18 Radiation fog is most likely to form:
(a) over a cool sea by night.
(b) over a warm sea by night.
(c) over land on cool, clear nights.
(d) over land during the afternoon.

19 Advection fog is most likely to form:
(a) over a cool sea by night.
(b) over a warm sea by night.
(c) over land on cool, clear nights.
(d) over land during the afternoon.
(e) at any time, day or night, when the conditions are right.

20 Is radiation fog formed as moist air loses heat to a cold land surface the result of an adiabatic process?

21 Is advection fog formed as warm moist air moves over a cold land mass the result of an adiabatic process?

22 When flying beneath an inversion, visibility is likely to be poor because of:
(a) mist, fog or smog.
(b) showers from cumulus clouds.

23 The figures '9999' in a meteorological forecast or report mean …

24 The visibility group '6000' in a meteorological forecast or report means …

25 The visibility group 'R35/0400' in a meteorological forecast or report means …

Exercises 26

Weather Forecasts and Reports

1 The primary method of obtaining meteorological information in the UK prior to flight (is/is not) by self-briefing using information in the aerodrome briefing area or by using the telephone recorded-message service.

2 A weather forecast is:
(a) a prediction.
(b) an observation.

3 A weather report is:
(a) a prediction.
(b) an observation.

4 The forecast of expected weather at an aerodrome is called an .... and it has the code name .... .

5 A routine aerodrome report has the code name M..... .

6 Significant weather that may affect a flight may be advised in the form of a ....

7 List four of the types of significant weather phenomena which may affect the safety of flight operations and that could be passed by the Flight Information Service to the pilot in the form of a SIGMET.

8 At aerodromes with suitable communications facilities such as AFTN or Telex, Area Forecasts plus Aerodrome Forecasts (TAFs) and/or Aerodrome Reports (METARs) (are/are not) available in text form.

9 AIRMET information is available:
(a) through the post.
(b) by radio.
(c) via the public telephone network (phone or fax).
(d) from aerodrome briefing offices connected to AFTN or Telex.

10 There (is/is not) a standard form or 'proforma' on which to note down the appropriate AIRMET information, when utilising the AIRMET telephone recording service.

11 The service in which meteorological reports and trends for a number of selected aerodromes are broadcast continuously on discrete VHF frequencies is called (METAR/TAF/ VOLMET/SIGMET/Area Forecast).

12 Weather information for certain aerodromes is available in recorded form on the VOLMET service:
(a) through the post.
(b) by radio.
(c) via the public telephone network.
(d) from ATC prior to flight.

13 A forecast for a particular aerodrome, usually issued for a 9-hour period, is called a (METAR/TAF/VOLMET/ SIGMET/area forecast).

14 A report of weather at a particular aerodrome is called a (METAR/TAF/ VOLMET/SIGMET/Area Forecast).

15 A visibility term in a TAF or a METAR of: '4000' means a visibility of ....

16 The term: 'NOSIG' appended to a METAR means .... expected for a period of .... hours after the time of the observation.

17 The cloud description: 'BKN035' in a TAF or METAR means between .... and .... oktas of cloud, with a base of ..... ft (amsl/aal).

18 Decode the following TAF for Guernsey:
EGJB 1221 23015KT 9999 BKN025=.

19 Decode the following TAF for Wick:
EGPC 1221 24015G25KT 6000 SCT035 TEMPO 5000 RASH BKN020=.

20 In TAFs and METARs, the cloud base is given as the height above:
(a) the aerodrome.
(b) mean sea level.
(c) the highest ground within 10 nm of the aerodrome.

21 In a TAF, '9999' means ...

22 What is meant by 'CAVOK'?

23 What is meant by 'TEMPO'?

24 What is meant by 'PROB 20' in a forecast?

25 What is meant by the term: 'RVR'?

26 What is meant by the term: 'TCU' appended to a cloud statement?

27 When 'EMBD' is mentioned in reference to thunderstorms in a forecast or report, what does it signify?

28 If the weather forecast for a particular aerodrome included the term 'BECMG', you (would/would not) expect a permanent change to the weather to become established during the forecast period.

29 In a TAF, the time group '1220' means …

30 In a METAR, the time group '1220' means …

31 A METAR finishes off 'Q1014 NOSIG='. What does the *Q1014* mean?

32 What is meant by '+' or '–' symbols prior to a 'met' phenomenon?

33 What is meant by the letters 'RADZ'?

34 What is meant by the term '+SHRAGR'?

35 The term 'NSW' in a TAF or METAR Trend means .. ……… ……… .

36 A temperature group '11/08' in a METAR means …

37 A temperature group '03/M01' in a METAR means …

38 The term 'SAUK' means a UK (TAF/METAR).

39 The term 'FCUK' means a UK (TAF/METAR).

40 The term 'TAF AMD' means ……. ……… ……… .

41 What does the following symbol mean?
⎾⚡

42 What does the following symbol mean?
∪⫴⟋

43 What does the following symbol mean?
△

44 What does the following symbol mean?
≡

45 What does the following symbol mean?
≡

46 What does the following symbol mean?
⸝∪

47 What does the following symbol mean?
_∧_ 200
    ⟍ 40

48 What is meant by the letters 'BR'?

49 What is meant by the letters 'FZ'?

50 What is meant by the letters 'LYR'?

51 What does the following symbol mean?
◄▲ ▲

Read AIP GEN 3-5 *Meteorological Charts – Symbology* and test yourself on its contents.

Meteorology

Answers 16

The Atmosphere

1 equator
2 troposphere
3 stratosphere, tropopause
4 oxygen, nitrogen, water vapour
5 (b)
6 decreases
7 (a)
8 decreases
9 true
10 (a)
11 more

Answers 17

Heating Effects in the Atmosphere

1 (b)
2 (a)
3 rises
4 sinks
5 (b)
6 (a)
7 less
8 less
9 cooler
10 warmer
11 higher
12 reduces
13 reduces
14 radiation
15 conduction
16 advection
17 convection
18 onshore
19 offshore
20 katabatic
21 anabatic
22 temperature inversion
23 50°F
24 5°F
25 5°C

26 increases

Answers 18

Atmospheric Pressure

1 decreases
2 1 mb(hPa) per 30 ft
3 semi-diurnal
4 (a)
5 pressure gradient
6 areas of high pressure to areas of low pressure
7 (a)
8 over-read, lower

Answers 19

The International Standard Atmosphere (ISA)

1 International Standard Atmosphere
2 +15°C
3 1013.2 mb(hPa)
4 1,225 grams per cubic metre
5 2°C
6 temperature lapse rate
7 −57°C
8 −3°C

Answers 20

Wind

1 wind
2 true north, °T
3 magnetic, °M
4 34 kt from 280°T
5 °M, 25 knots
6 veered
7 backed
8 veered
9 pressure gradient force
10 Coriolis force
11 rotation, poles
12 greater
13 right

14 geostrophic wind
15 left
16 refer to the text
17 low
18 high
19 right drift, altimeter will read too high
20 exceed
21 be less than
22 (c)
23 10 metres
24 decreased, friction
25 back
26 land
27 day
28 more
29 stronger
30 more
31 (d)
32 in the vicinity of (and perhaps up to 10 nm distance from) the actual cloud
33 (a)
34 windshear
35 is possible
36 lee
37 mountain waves, standing waves
38 east
39 headwind
40 lenticular
41 lee
42 40 nm

Answers 21

Cloud

1 cirriform, cumuliform, stratiform and nimbus
2 cumuliform
3 stratiform
4 cumulus
5 cumulonimbus
6 nimbus
7 gives off
8 humidity
9 more
10 higher
11 decreases
12 saturated

13 relative humidity
14 (a)
15 increases
16 (b)
17 (d)
18 higher, closer to
19 condense out as liquid water
20 cools
21 adiabatic
22 increases
23 yes
24 3°C/1,000 ft, dry adiabatic lapse rate
25 cloud
26 gives off, decreases
27 one-half, 1.5°C/1,000 ft
28 environmental lapse rate
29 (a)
30 3,000 ft agl
31 4,000 ft agl
32 5,000 ft agl
33 3,000 ft agl
34 turbulence cloud
35 orographic cloud
36 cumuliform
37 stratiform
38 warmer, drier, foehn wind effect
39 3°C/1,000 ft unsaturated, 1.5°C/1,000 ft saturated
40 rain
41 hail
42 snow
43 cumuliform
44 stratiform
45 virga
46 lenticular

Answers 22

Thunderstorms

1 instability, moisture and a trigger action
2 the cumulus stage, the mature stage, the dissipating stage
3 refer to the text
4 the early cumulus stage
5 higher than
6 the mature stage
7 will be, cool downdrafts

8   anvil
9   False
10 (c)
11 refer to the text

Answers 23

Air Masses and Frontal Weather

1   maritime
2   continental
3   unstable
4   stable
5   stable and moist
6   subsidence
7   convergence
8   low
9   high
10 warmer, drier and more stable
11 stability
12 instability
13 instability
14 warm front
15 lowers
16 stratiform
17 cold
18 precedes
19 1 in 150
20 cirrus, 600 nm
21 (c)
22 refer to the text
23 cold front
24 cumuliform
25 may
26 showers
27 veer
28 fall, fall
29 (c)
30 1 in 50
31 fall
32 veer
33 rise
34 rise, rise
35 north
36 true
37 better
38 occluded front
39 may

40 (d)
41 anticlockwise
42 right (starboard)
43 clockwise
44 left (port)
45 convergence
46 rising
47 trough
48 faster
49 weaker
50 divergence
51 subsiding
52 stable
53 anti-cyclone
54 ridge
55 col
56 warm
57 descend

Answers 24

Icing

1   0°C
2   airframe ice
3   carburettor, can
4   (a)
5   may
6   supercooled
7   (a)
8   (d)
9   (b)
10 is
11 true
12 (b)
13 (a)
14 (c)
15 (b)
16 low
17 (c)

Answers 25

Visibility

1   slant
2   fog
3   mist
4   radiation fog
5   cloudless night, moist air, light winds

6 advection fog
7 advection fog
8 frontal fog
9 inversion
10 dawn
11 (b)
12 (a)
13 poorer
14 (a)
15 (b)
16 (d) or (c)
17 (c)
18 (c)
19 (e)
20 no, since there is a transfer of heat from the air mass to the land
21 no, since there is a transfer of heat from the air mass to the land
22 (a)
23 visibility of 10 km or greater
24 visibility of 6,000 metres (6km)
25 Runway Visual Range of 400 metres on Runway 35

Answers 26

Weather Forecasts and Reports

1 is
2 (a)
3 (b)
4 Aerodrome Forecast, TAF
5 METAR
6 SIGMET
7 active thunderstorms, tropical revolving storms, a severe line squall, heavy hail, severe turbulence, severe airframe icing, marked mountain waves, widespread dust or sandstorm
8 are
9 (c) and (d)
10 is
11 VOLMET
12 (b)
13 TAF
14 METAR
15 4,000 metres
16 no significant change, 2 hours

17 5–7 oktas, at base 3,500 ft aal
18 from 1200–2100 UTC, wind 230°/15 kt, visibility in excess of 10 km, 5–7 oktas of cloud at base 2500 ft aal
19 1200–2100 UTC, wind 240°/15 kt, with gusts to 25 kt, visibility 6,000 metres, 3–4 oktas of cloud at base 3500 ft aal, with temporary deteriorations (less than 60 minutes) visibility 5,000 metres in moderate rain showers, and cloud 5–7 oktas at 2,000 ft aal
20 (a)
21 visibility in excess of 10 km
22 refer to the text
23 refer to the text
24 20% probability
25 runway visual range
26 the cloud is towering cumulus
27 the thunderstorms are embedded in other cloud (and are therefore difficult to see)
28 would
29 from 1200 to 2000 UTC
30 at time 1220 UTC
31 QNH is 1014 mb
32 + = heavy; – = light
33 rain and drizzle
34 heavy rain showers with hail
35 no significant weather
36 observed air temperature +11°C; dewpoint temperature +8°C
37 observed air temperature +3°C; dewpoint temperature −1°C
38 METAR
39 TAF
40 amended Aerodrome Forecast
41 thunderstorm
42 severe aircraft icing
43 hail
44 widespread mist
45 widespread fog
46 freezing precipitation
47 severe turbulence from 4,000 ft to 20,000 ft amsl
48 mist
49 freezing
50 layer or layered
51 cold front at the surface

JAR-FCL Abbreviations

Definitions for JAR-FCL terms begin on page 126. Definitions for ICAO terms used in the JARs begin on page 173.

A	Aeroplane
A/C	Aircraft
AMC	Acceptable Means of Compliance; Aeromedical Centre
AME	Authorised Medical Examiner
AMS	Aeromedical Section
ATC	Air Traffic Control
ATP	Air Transport Pilot
ATPL	Air Transport Pilot Licence
CFI	Chief Flying Instructor
CGI	Chief Ground Instructor
CPL	Commercial Pilot Licence
CRE	Class Rating Examiner
CRI	Class Rating Instructor
CQB	Central Question Bank
FCL	Flight Crew Licensing
FE	Flight Examiner
FI	Flight Instructor
FIE	Flight Instructor Examiner
FNPT	Flight and Navigation Procedures Trainer
FS	Flight Simulator
FTD	Flight Training Device
FTO	Flight Training Organisation
H	Helicopter
HT	Head of Training
ICAO	International Civil Aviation Organization
IEM	Interpretative and Explanatory Material
IFR	Instrument Flight Rules
IMC	Instrument Meteorological Conditions
IR	Instrument Rating
IRE	Instrument Rating Examiner
IRI	Instrument Rating Instructor
JAA	Joint Aviation Authorities

JAR	Joint Aviation Requirements
LOFT	Line Orientated Flight Training
MCC	Multi Crew Cooperation
ME	Multi-engine
MEP	Multi-engine Piston
MET	Multi-engine Turboprop
MPA	Multi-pilot Aeroplane
MPH	Multi-pilot Helicopter
NM	Nautical Miles
OML	Operational Multicrew Limitation
OSL	Operational Safety Pilot Limitation
OTD	Other Training Devices
PF	Pilot Flying
PIC	Pilot-in-Command
PICUS	Pilot-in-Command Under Supervision
PNF	Pilot Not Flying
PPL	Private Pilot Licence
R/T	Radiotelephony
SE	Single-engine
SEP	Single-engine Piston (Aeroplanes)
SET	Single-engine Turboprop
SFE	Synthetic Flight Examiner
SFI	Synthetic Flight Instructor
SIM	Simulator
SPA	Single-pilot Aeroplane
SPH	Single-pilot Helicopter
SPIC	Student Pilot-in-Command
STD	Synthetic Training Devices
TMG	Touring Motor Glider
TR	Type Rating
TRE	Type Rating Examiner
TRI	Type Rating Instructor
TRTO	Type Rating Training Organisation
VFR	Visual Flight Rules
VMC	Visual Meteorological Conditions

Index